Polypropylene - Polymerization and Characterization of Mechanical and Thermal Properties

Edited by Weiyu Wang and Yiming Zeng

Published in London, United Kingdom

IntechOpen

Supporting open minds since 2005

Polypropylene - Polymerization and Characterization of Mechanical and Thermal Properties
http://dx.doi.org/10.5772/intechopen.73995
Edited by Weiyu Wang and Yiming Zeng

Contributors
Antonella Patti, Domenico Acierno, Luca Fambri, Luca Lutterotti, Koh-Hei Nitta, Xiong Wang, Xiaoyu Han, Renwei Xu, Valeria Pettarin, Alejandra Costantino, Caren Rosales, Weiyu Wang

First published in London, United Kingdom, 2020 by IntechOpen
IntechOpen is the global imprint of INTECHOPEN LIMITED, registered in England and Wales, registration number: 11086078, 7th floor, 10 Lower Thames Street, London, EC3R 6AF, United Kingdom
Printed in Croatia

British Library Cataloguing-in-Publication Data
A catalogue record for this book is available from the British Library

Additional hard and PDF copies can be obtained from orders@intechopen.com

Polypropylene - Polymerization and Characterization of Mechanical and Thermal Properties
Edited by Weiyu Wang and Yiming Zeng
p. cm.
Print ISBN 978-1-83880-414-5
Online ISBN 978-1-83880-416-9
eBook (PDF) ISBN 978-1-83880-429-9

For EU product safety concerns:
IN TECH d.o.o., Prolaz Marije Krucifikse Kozulić 3, 51000 Rijeka, Croatia, info@intechopen.com or visit our website at intechopen.com.

Meet the editors

Dr. Weiyu Wang graduated from the University of Tennessee, Knoxville, in 2015 with a PhD in Polymer Chemistry. After working as a post-doc research associate in Oak Ridge National Lab, Dr. Wang joined the South China Advanced Institute for Soft Matter Science and Technology as a material consultant. Currently, he is pursuing an MBA at Yale School of Management. As a chemist, Dr. Wang's primary research focus is to understand the fundamental principles behind the structure–property relationship of polymeric materials and develop advanced materials for applications in high-performance elastomers, energy storage, and electronic and medical devices. One of his career passions/interests is to bring together universities and small- and middle-scale industries to facilitate the advancement in green technologies.

Dr. Yiming Zeng graduated from Tsinghua University, China, in 2012 and was awarded a BSc in Chemical Engineering in 2012. He then joined the University of Minnesota, Twin Cities, and graduated in 2018 with a PhD in Chemical Engineering. He was co-advised by Dr. Frank Bates and Dr. Timothy Lodge. Upon graduation, Dr. Zeng joined the H.B. Fuller Company, Minneapolis, as a senior scientist. He is now active in formulating the next generation of adhesives for the hygiene industry, with the intent to optimize the performance–cost balance of adhesives. His interest lies in understanding the structure–property relationship of blend systems containing polyolefins, and utilizing this knowledge to expand the application of polyolefins in the industry.

Contents

Preface XI

Chapter 1 1
Introductory Chapter: Polypropylene - Synthesis and Functionalization
by Weiyu Wang

Chapter 2 9
Versatile Propylene-Based Polyolefins with Tunable Molecular Structure through Tailor-Made Catalysts and Polymerization Process
by Xiong Wang, Xiaoyu Han and Renwei Xu

Chapter 3 37
Thermal Conductivity of Polypropylene-Based Materials
by Antonella Patti and Domenico Acierno

Chapter 4 57
Polypropylene Blends and Composite: Processing-Morphology-Performance Relationship of Injected Pieces
by Maria Alejandra Costantino, Caren Rosales and Valeria Pettarin

Chapter 5 71
Tensile Properties in β-Modified Isotactic Polypropylene
by Koh-hei Nitta and Tsutomu Takashima

Chapter 6 91
Effect of Processing and Orientation on Structural and Mechanical Properties of Polypropylene Products
by Luca Fambri and Luca Lutterotti

Preface

Polypropylene (PP) is one of the most important commodity plastics that was first synthesized by J. Paul Hogan and Robert L. Banks in 1951. After almost seven decades of innovation in science and engineering, PP-based materials have found tremendous applications in automobiles, industrial moldings, consumer packaging, and medical devices. The objective of this book is to provide an overview of the progress in PP from the perspectives of synthesis, structure–property relationship, processing, PP composites, and applications.

In the chapter Versatile Propylene-Based Polyolefins with Tunable Molecular Structure through Tailor-Made Catalysts and Polymerization Process, Dr. Xiong Wang reviews the chemical structure, catalytic systems, and various polymerization processes to synthesize PP.

In the chapter Thermal Conductivity of Polypropylene-Based Materials, Dr. Antonella Patti addresses the growing demand in thermal conductive plastics by providing an overview of the thermal conductivity of PP and methods to enhance the thermal conductivity of PP-based composites.

In the chapter Polypropylene Blends and Composite: Processing-Morphology-Performance Relationship of Injected Pieces, Dr. Alejandra Costantino summarizes the progress and fire retardancy mechanism of PP composite-based nano-fire retardants.

In the chapter Tensile Properties in β-Modified Isotactic Polypropylene, Dr. Koh-Hei Nitta prepares spherulitic isotactic PPs by adding β-nucleators and investigates the effects the β-phase modification on the mechanical properties.

In the last chapter, Effect of Processing and Orientation on Structural and Mechanical Properties of Polypropylene Products, Dr. Luca Fambri illustrates the mechanical and structural properties of oriented i-PP homopolymers by using various characterization methods.

Dr. Weiyu Wang
South China University of Technology,
Guangzhou, China

Dr. Yiming Zeng
H.B. Fuller Company,
United States

Chapter 1

Introductory Chapter: Polypropylene - Synthesis and Functionalization

Weiyu Wang

1. Introduction

First discovered by J. Paul Hogan and Robert L. Banks in 1951, polypropylene (PP) is polymerized from propylene out of crude oil and is the most widely used commodity thermoplastic by volume [1]. Over the past 70 years, significant progress has been achieved to manufacture and commercialize PP (**Figure 1**) [2, 3]. The most recent milestone in the field of polypropylene is from PureCycle Technology, where the waste carpet has been successfully purified into clear, odorless ultrapure recycled polypropylene (UPRP) resin [4]. Through not fully commercialized yet, the innovation opens up a new venue of recycling processed PP into raw materials as a resin.

As one of the cheapest plastics with great processability, chemical resistance, and moisture barriers, PP with different tacticity found various downstream applications in textile, automotive, cosmetics, and consumer packaging. In 2016, 26% of polymer demand in the world by volume was from PP (**Figure 2**) [5].

2. Synthesis and functionalization

Well-defined polymers with narrow polydispersity and controlled molecular weight are essential to delineate the structure–property relationship of polymeric materials [6]. Using N,N-diethyl hafnium derivative as active transition metal propagation center and $ZnEt_2$ as metal alkyl chain transfer agent, Sita first demonstrated the living coordinative chain transfer polymerization of propene to produce amorphous atactic polypropylene (a-PP) with narrow polydispersity and various molecular weights from 12.6 kDa to 111 kDa [7]. Compared with a-PP, isotactic polypropylene (iPP) is more practical for applications in packaging and automotive parts. Coates designed a pyridylamidohafnium catalyst that produced PP with high iso-selectivity (91%) and living polymerization behavior (**Figure 3**) [8]. The study also confirmed a ligand-monomer interaction as the mechanism of stereo-control. The progress of using coordinative chain transfer polymerization has been thoroughly reviewed elsewhere [9, 10].

Block copolymers have contributed significantly to thermoplastic elastomers, soft lithography, and drug delivery [11]. Block copolymers containing polypropylene can serve as a compatibilizer to improve the interface interaction between polyolefin and other polar materials. Chen demonstrated an early example of polypropylene-block-poly(methyl methacrylate) (PP-b-PMMA) diblock copolymer using Group IV metal catalyst [12]. By taking advantage of the solubility, the

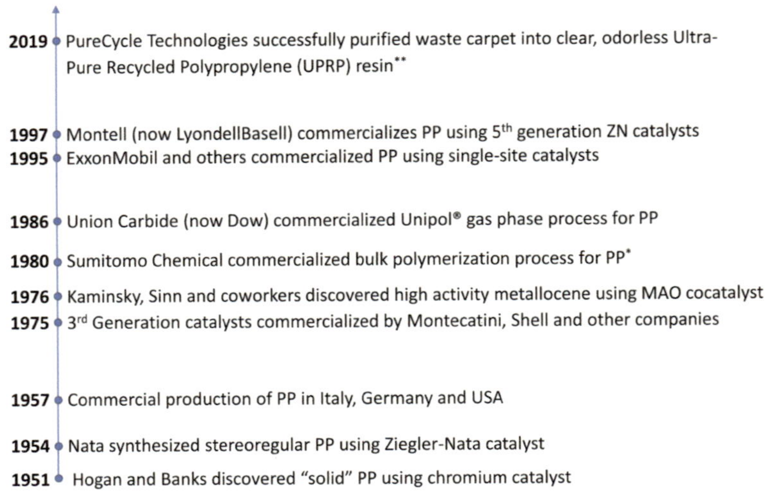

Figure 1.
Key milestones of the commercialization of polypropylene. Adapted from Ref. [1-4].

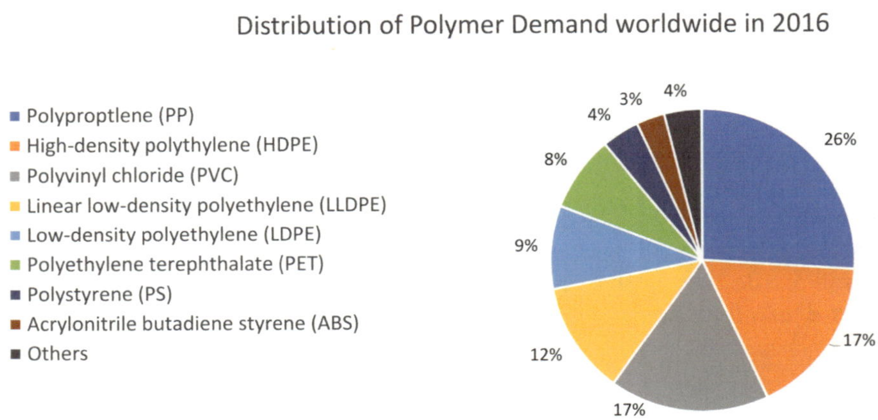

Figure 2.
Distribution of polymer demand worldwide in 2016 by volume. Adapted from Ref. [5].

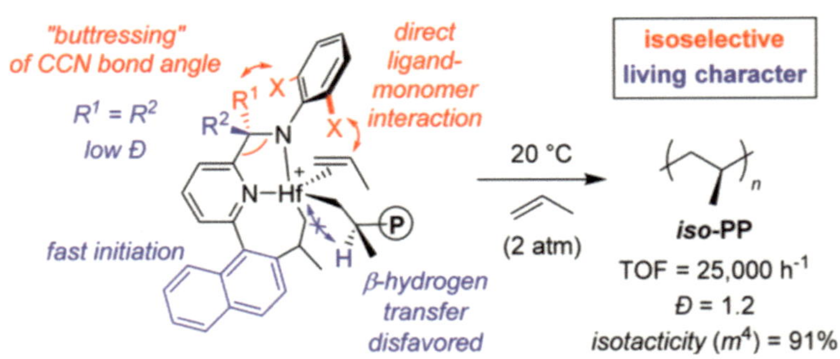

Figure 3.
Living and iso-selective propylene polymerization (reprinted with permission from Ref. [8]. Copyright 2017 American Chemical Society).

resulting polymer could be purified into narrow distributed copolymers with hexane-heptane fractionation. Dong et al. used a brominated isotactic PP with styrene termination as a macroinitiator to copolymerize styrene or methyl

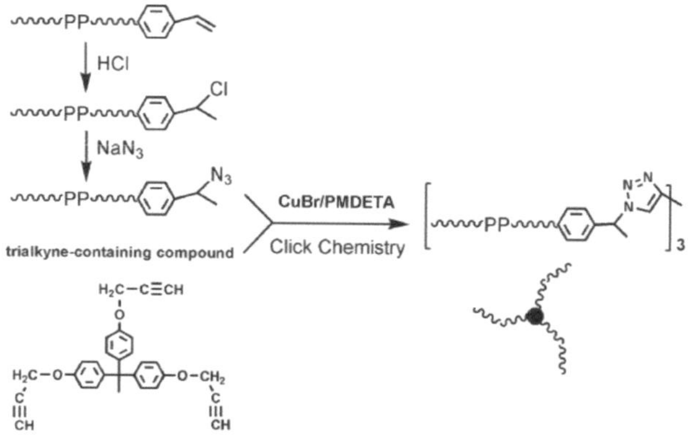

Figure 4.
Synthesis of three-arm PP star polymer by click chemistry (reprinted with permission from Ref. [14]. Copyright 2010 American Chemical Society).

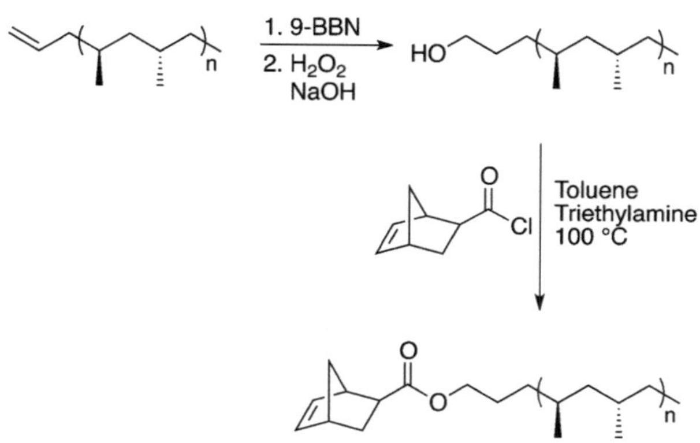

Figure 5.
Synthesis of PP comb polymers (reprinted with permission from Ref. [20]. Copyright 2012 American Chemical Society).

methacrylate and studied the blending behavior between the resulting copolymers with PS/PP or PMMA/PP blend [13]. The same group prepared azide end-functionalized PP and prepared PP three-arm star polymer (**Figure 4**) [14] and PP-b-polycaprolactone (PP-b-PCL) block copolymers [15]. By terminating the coordination polymerization of iPP with (p-vinylphenyl)chlorosilane, iPP star polymers with 3–8 different numbers of arms were prepared in an H_2O/toluene emulsion system [16].

Using stereoselective insertion polymerization catalyst, Coates synthesized a group of block copolymers containing iPP as the rigid block and regioirregular polypropylene (rPP) as the elastic block. The mechanical characterization of the iPP-rPP-iPP-rPP-iPP pentablock copolymer indicated a strain at break of 2400% and a maximum true tensile stress of 250 MPa [17]. In collaboration with Bates and LaPointe, the group synthesized a polyethylene-b-iPP-b-polyethylene-b-iPP tetrablock copolymer and evaluated the blending behavior of the tetrablock copolymer with PE/iPP. By "welding" polyethylene (PE) and iPP, together with the tetrablock copolymer, previously impossible due to the immiscibility, the blend was transformed from a brittle glass into a tough plastic, paving a possibility to recycle the world's two most-produced polymer materials [18].

Long-chain branching in polymers has shown interesting rheological behaviors [19]. Using norbornene-terminated syndiotactic PP as a macromonomer, Coates synthesized a group of well-defined s-PP bottlebrush polymers with a molecular weight from 46 kDa to 172 kDa using Grubbs' metathesis catalyst (**Figure 5**) [20]. A decrease in both melting and crystallization temperature was observed and attributed to the constraints on the rigid backbone. Further research by Bates and Hillmyer revealed a scaling transition that depends on the length of the backbone [21]. Hazer evaluated the surface property and mechanical property of a group of graft copolymer containing polypropylene as the backbone and polyethylene glycol (PEG) as the side chain [22]. With 15% of PEG, the graft copolymer demonstrated ultimate stress of 22 MPa and elongation at break of 670%. Bielawski developed a direct C—H azidation method to introduce azide functionalities into commercially available PP [23] and prepared PP-g-PEG using click chemistry. Tasdelen used a similar approach and synthesized PP-g-PCL copolymers [24].

"Reactive" polyolefin approach, adding functional monomer units into the polyolefin chain, has emerged as a powerful tool to chemically functionalize polyolefins [25]. Pan copolymerized p-(3-butenyl)styrene and propylene with (pyridylamido)Hf/[Ph_3C][$B(C_6F_5)_4$]/Al^iBu_3 catalytic system, which selectively copolymerize α-olefin over styrene [26]. The pendant styrenic vinyl groups in the resulting polymer were quantitatively converted into carboxylic acid groups with thiol-ene addition. The same group further extended this methodology to prepare amino-containing iPP, which exhibited high thermal stability and melting temperature [27]. Chung synthesized a group of hydroxyl-functionalized PP using silane-protected 10-undecen-1-ol as a comonomer and converted the hydroxyl pendant group into butylated hydroxytoluene (BHT) derivatives (**Figure 6**) [28]. The resulting BHT-functionalized PP demonstrated improved thermal stability and higher dielectric constant. With two methylene group spacers between BHT and ester linkage, the materials displayed superior thermal stability at 190°C compared with general and capacitor grade PP [29].

Polypropylene is one of the most important plastics in our daily life. However, the materials itself also caused a significant amount of plastic pollution. As much exciting progress has been achieved recently to introduce functionalities and improve both mechanical and thermal stabilities, the research community should also emphasize on developing approaches to recycle PP and PE from the processed product.

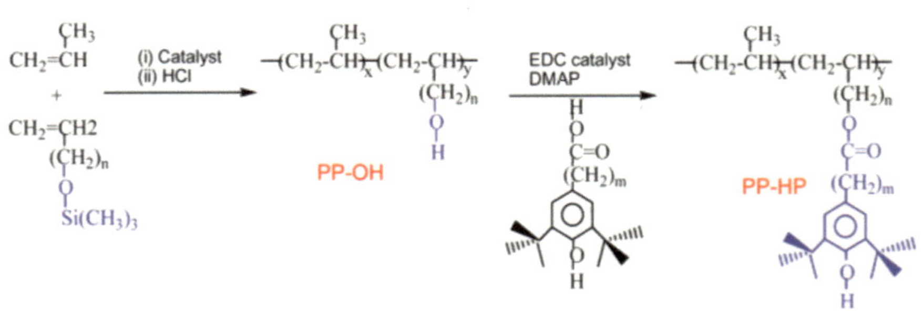

Figure 6.
Synthesis of BHT-functionalized PP (reprinted with permission from Ref. [28]. Copyright 2015 American Chemical Society).

Author details

Weiyu Wang
South China Advanced Institute for Soft Matter Science and Technology,
South China University of Technology, Guangzhou, P.R. China

*Address all correspondence to: wwang41@vols.utk.edu

References

[1] Stinson S. Discoverers of polypropylene share prize. Chemical & Engineering News. 1987;**65**(10):30. DOI: 10.1021/cen-v065n010.p030

[2] Dennis B. Malpass, Elliot I. Band. Introduction to polymers of propylene. In: Dennis B. Malpass, Elliot I. Band, editors. Introduction to Industrial Polypropylene: Properties, Catalysts Processes. 1st ed. John Wiley & Sons; 2012. p. 1-18. DOI: 10.1002/9781118463215

[3] Sato H, Ogawa H. Review on Development of Polypropylene Manufacturing Process. Vol. 12. Sumitomo Chemical Co., Ltda, Process and Production Technology Center. Tokyo, Japan: Fecha de consulta; 2009

[4] PureCycle Technologies. PureCycle Technologies Celebrates Successful Run of Groundbreaking Plastics Recycling Technology [Internet]. September 2019. [Updated: September 2019]. Available from: https://purecycletech.com/2019/09/successful-run-of-feedstock-evaluation-unit/ [Accessed: March 2020]

[5] Statista. Plastic industry worldwide [Internet]. January 2020. [Updated: January 2020]. Available from: https://www.statista.com/study/51465/global-plastics-industry/ [Accessed: March 2020]

[6] Lutz J-F. 100th anniversary of macromolecular science viewpoint: Toward artificial life-supporting macromolecules. ACS Macro Letters. 2020;**9**:185-189. DOI: 10.1021/acsmacrolett.9b00938

[7] Zhang W, Sita LR. Highly efficient, living coordinative chain-transfer polymerization of propene with $ZnEt_2$: Practical production of ultrahigh to very low molecular weight amorphous atactic polypropenes of extremely narrow polydispersity. Journal of the American Chemical Society. 2008;**130**(2):442-443. DOI: 10.1021/ja078120v

[8] Domski GJ, Eagan JM, De Rosa C, Di Girolamo R, LaPointe AM, Lobkovsky EB, et al. Combined experimental and theoretical approach for living and Isoselective propylene polymerization. ACS Catalysis. 2017;7(10):6930-6937. DOI: 10.1021/acscatal.7b02107

[9] Valente A, Mortreux A, Visseaux M, Zinck P. Coordinative chain transfer polymerization. Chemical Reviews. 2013;**113**(5):3836-3857. DOI: 10.1021/cr300289z

[10] Stürzel M, Mihan S, Mülhaupt R. From multisite polymerization catalysis to sustainable materials and all-polyolefin composites. Chemical Reviews. 2016;**116**(3):1398-1433. DOI: 10.1021/acs.chemrev.5b00310

[11] Feng H, Lu X, Wang W, Kang NG, Mays JW. Block copolymers: Synthesis, self-assembly, and applications. Polymers. 2017;**9**(10):494. DOI: 10.3390/polym9100494

[12] Jin J, Chen EY-X. Stereoblock copolymerization of propylene and methyl methacrylate with single-site metallocene catalysts. Macromolecular Chemistry and Physics. 2002;**203**(16): 2329-2333. DOI: 10.1002/macp.200290010

[13] Zhang L-Y, Fan G-Q, Guo C-Y, Dong J-Y, Hu Y-L, Huang M-B. Synthesis of polypropylene block copolymers from brominated styrene-terminated isotactic polypropylene. European Polymer Journal. 2006;**42**(5):1043-1050. DOI: 10.1016/j.eurpolymj.2005.11.021

[14] Huang H, Niu H, Dong J-Y. Synthesis of star isotactic polypropylene using click chemistry. Macromolecules. 2010;**43**(20): 8331-8335. DOI: 10.1021/ma1019335

[15] Huang H, Niu H, Dong J-Y. Synthesis of azide end-functionalized

isotactic polypropylene building block and renewed modular synthesis of diblock copolymers of isotactic polypropylene and poly (ε-caprolactone). Journal of Polymer Science Part A: Polymer Chemistry. 2011;**49**(10):2222-2232. DOI: 10.1002/pola.24653

[16] Liu X, Niu H, Yang L, Dong J-Y. New effort to synthesize star isotactic polypropylene. Polymer Chemistry. 2018;**9**(24):3347-3354. DOI: 10.1039/C8PY00318A

[17] Hotta A, Cochran E, Ruokolainen J, Khanna V, Fredrickson GH, Kramer EJ, et al. Semicrystalline thermoplastic elastomeric polyolefins: Advances through catalyst development and macromolecular design. Proceedings of the National Academy of Sciences. 2006;**103**(42):15327-15332. DOI: 10.1073/pnas.0602894103

[18] Eagan JM, Xu J, Girolamo RD, Thurber CM, Macosko CW, AMLaPointe FSB, et al. Combining polyethylene and polypropylene: Enhanced performance with PE/iPP multiblock polymers. Science. 2017;**355**(6327):814-816. DOI: 10.1126/science.aah5744

[19] Lagendijk RP, Hogt AH, Buijtenhuijs A, Gotsis AD. Peroxydicarbonate modification of polypropylene and extensional flow properties. Polymer. 2001;**42**(25):10035-10043. DOI: 10.1016/S0032-3861(01)00553-5

[20] Anderson-Wile AM, Coates GW, Auriemma F, De Rosa C, Silvestre A. Synthesis and ring-opening metathesis polymerization of norbornene-terminated syndiotactic polypropylene. Macromolecules. 2012;**45**(19):7863-7877. DOI: 10.1021/ma301073s

[21] Dalsin SJ, Hillmyer MA, Bates FS. Molecular weight dependence of zero-shear viscosity in atactic polypropylene bottlebrush polymers. ACS Macro Letters. 2014;**3**(5):423-427. DOI: 10.1021/mz500082h

[22] Balcı M, Allı A, Hazer B, Güven O, Cavicchi K, Cakmak M. Synthesis and characterization of novel comb-type amphiphilic graft copolymers containing polypropylene and polyethylene glycol. Polymer Bulletin. 2010;**64**(7):691-705. DOI: 10.1007/s00289-009-0211-3

[23] Liu D, Bielawski CW. Direct azidation of isotactic polypropylene and synthesis of 'grafted to' derivatives thereof using azide–alkyne cycloaddition chemistry. Polymer International. 2017;**66**(1):70-76. DOI: 10.1002/pi.5180

[24] Acik G, Sey E, Tasdelen MA. Polypropylene-based graft copolymers via CuAAC click chemistry. eXPRESS Polymer Letters. 2018;**12**(5):418-428. DOI: 10.3144/expresspolymlett.2018.35

[25] Chung TCM. Functional polyolefins for energy applications. Macromolecules. 2013;**46**(17):6671-6698. DOI: 10.1021/ma401244t

[26] Wang X-Y, Wang Y-X, Li Y-S, Pan L. Convenient syntheses and versatile functionalizations of isotactic polypropylene containing plentiful pendant styrene groups with high efficiency. Macromolecules. 2015;**48**(7):1991-1998. DOI: 10.1021/acs.macromol.5b00128

[27] Shang R, Gao H, Luo F, Li Y, Wang B, Ma Z, et al. Functional isotactic polypropylenes via efficient direct copolymerizations of propylene with various amino-functionalized α-olefins. Macromolecules. 2019;**52**(23):9280-9290. DOI: 10.1021/acs.macromol.9b00757

[28] Zhang G, Li H, Antensteiner M, Mike Chung TC. Synthesis of functional polypropylene containing hindered

phenol stabilizers and applications in metallized polymer film capacitors. Macromolecules. 2015;**48**(9):2925-2934. DOI: 10.1021/acs.macromol.5b00439

[29] Zhang G, Changwoo N, Mike Chung TC, Petersson L, Hillborg H. Polypropylene copolymer containing cross-linkable antioxidant moieties with long-term stability under elevated temperature conditions. Macromolecules. 2017;**50**(18): 7041-7051. DOI: 10.1021/acs.macromol.7b01235

Chapter 2

Versatile Propylene-Based Polyolefins with Tunable Molecular Structure through Tailor-Made Catalysts and Polymerization Process

Xiong Wang, Xiaoyu Han and Renwei Xu

Abstract

Since the discovery of Ziegler-Natta catalysts for olefin polymerization in the 1950s, the production of polyolefins with a variety of properties has continuously grown with rapid development of catalyst technology combined with polymerization process innovation. For propylene-based polyolefin, various polyolefins with distinctive characteristic of mechanical and optical properties were made with specific catalysts in commercial industries owned especially by those large worldwide companies. In this chapter, Ziegler-Natta catalysts, metallocene catalysts, and post-metallocene catalysts for PP polymerization are discussed in detail. Gas phase, bulk, slurry, and solution polymerization processes, such as Spheripol (Basell), Hypol (Mitsui Chemicals), Unipol (Dow Chemical), Innovene (INEOS), Novelen (BASF), Spherizone (Basell), and Borstar (Borealis), developed by the industrial tycoons were reviewed. The molecular architecture of the PP-based polyolefins could be tailored precisely using specific high-performance catalyst in an appropriate polymerization process, and different types of PPs, including homopolypropylene (HPP), random copolypropylene (RPP), impact PP, PP-based block copolymer, functionalized PP, etc., are produced. The relationship between molecular structure and performance of the PP-based polyolefins is also discussed thereof.

Keywords: polymerization process, propylene-based polyolefin, polypropylene, Z-N catalyst, metallocene, post-metallocene

1. Introduction

The consumption of polyolefins has been remaining growing with continuous catalyst technology innovation since the discovery of Ziegler-Natta catalysts in the 1950s [1, 2]. Numerous technologies are adopted to improve the performance of regular homopolypropylene (HPP), such as toughness, tensile strength, and transparency, and a series of PP-based polyolefins including isotacticity polypropylene (iPP), random copolypropylene (RPP), impact PP, PP-based block copolymer, functionalized PP, etc. are successfully commercialized by tailor-made catalysts and polymerization process [3–8]. Now these PP-based polyolefins are used in a wide range of industries such as packaging, electrical and electronics, construction, automobile, medical, equipment, and facilities industries [9, 10].

According to the statistics from IHS Chemical (2018), the total production of polypropylene worldwide was about 56 million tons in 2016, and by 2022, about 75 million tons is predicted. The biggest increases have been taking place in Asia in recent years due to their dramatic expansion of economic share with a huge supply of cheap raw materials from Gulf Coast countries. China now possesses the largest market share in PP production of above 22 million tons.

Along with this massive production was the successful development of the catalyst technology and the polymerization process innovation [11]. The Z-N catalysts were at first immobilized in a carrier such as $TiCl_3$ [9]. The two most significant improvements were the evolutionary use of $MgCl_2$ as a reactive catalyst support, which can dramatically improve the catalyst performance with excellent shape control, and the discovery of electron donors (internal electron donors and external electron donors) in the catalyst system, which can improve the catalyst activity and control the stereoregularity, and led to the dramatic growth of iPP production.

Different from the multisite Z-N catalysts, single site metallocenes have not brought much attention to the polyolefin industry until the discovery of methylaluminoxane (MAO) by Sinn and Kaminsky due to the dramatic increase of polymerization activity as a cocatalyst. Typically, single-site metallocene catalysts make it possible to fine-tune the microstructure of the produced polymer chain by ligand design in the catalyst complexes, with excellent α-olefin incorporation ability. In 1990s, two successful samples of commercialization of metallocene catalysts were realized by ExxonMobil and Dow. In 1991, ExxonMobil was the first company to put metallocene catalysts into commercial use with the new Exxpol® Technology. Then in 1992, Dow launched the constrained geometry catalysts (CGC) with linked half-titanocenes containing amide ligands, which are still called metallocene catalysts for convenience, to produce metallocene-based polyolefins with INSITE™ technology [12].

Compared to the Exxpol® Technology developed by ExxonMobil, which is based on heterogeneous catalysts and gas-phase process, the INSITE® technology from Dow is based on the constrained geometry catalyst (CGC) in a solution process. The homogeneous catalyst system has the ability to control polymer microstructure with flexibility and simplicity from the homogeneous phase system. And also the relationship between the catalyst structure and the physical properties is easy to be characterized and modeled.

Polyolefin production from metallocene-based catalysts and a solution process were rapidly adopted for many applications; however, some drawbacks such as poor compression set and poor scratch resistance limited their applications. In 2004, a new post-metallocene catalyst with the pyridyl amine system was commercialized in the solution process by Dow to produce a family of propylene/ethylene copolymers called VERSIFY™ Plastomers and Elastomers [13, 14]. This pyridyl amine-based catalyst was developed through high throughput screening technology, and was suitable for production of propylene-based copolymers with high molecular weight over a wide range of chemical composition distributions.

In 2006, Dow announced olefin block copolymers which were produced by chain shuttling polymerization technology in a solution process. As shown in **Figure 1**, this technology employs two catalysts and a chain shuttling agent, and the two catalysts have totally different incorporation ability of α-olefin, thus producing different chain block-soft and hard PE segments-by chain shuttling agent (diethyl zinc), and the produced chains are composed of at least two alternating soft and hard segments [15]. The chain shuttling polymerization is illustrated in **Figure 3**. By combining the pyridyl-amine catalyst and CGC catalyst and alkylaluminum as the chain shuttling agent, the chain shuttling catalyst technology was also used to produce propylene-based stereoblock copolymer with a high molecular weight, at least some of which differ in irregular branching content, especially regio-irregular 2,1- and/or 3,1-monomer insertions [17], (**Figure 2**). Thus, the block copolymers obtain desirable properties due to the presence of alternating "soft" and "hard" blocks in the same polymer chain [11].

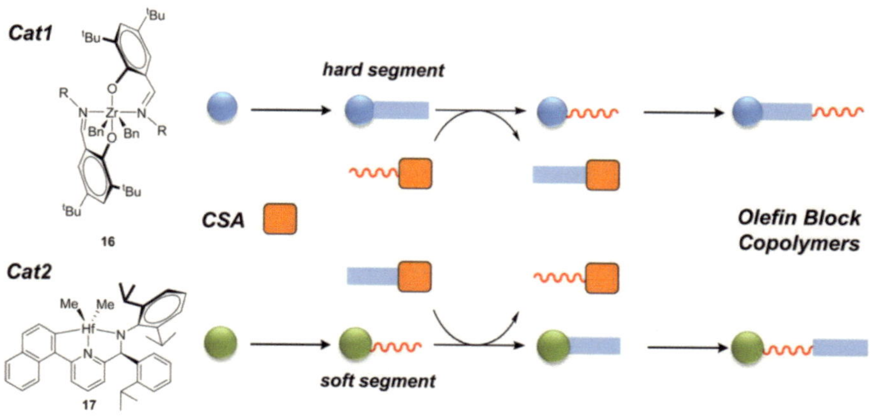

Figure 1.
Schematic illustration of chain shuttling polymerization [3].

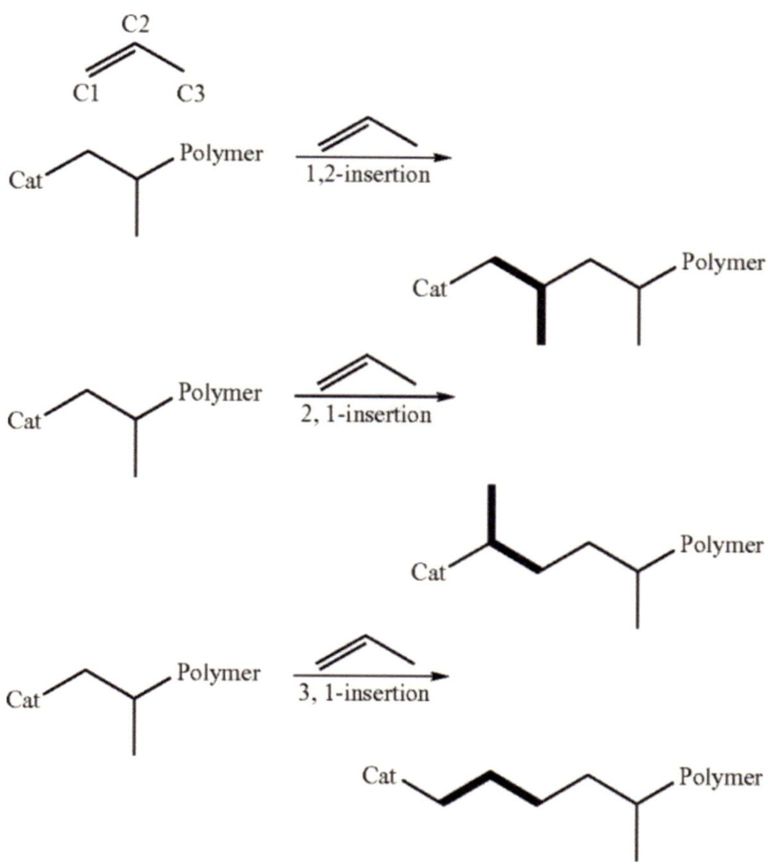

Figure 2.
Regio-irregular 2,1- and/or 3,1-monomer insertions [11].

Furthermore, numerous methods such as using polymerizable chain shuttling agents and reactive comonomers were adopted to prepare functionalized PP polymer [18] in order to improve PP's interactive performance and broaden its applications to high value-added products, in which compatibility and adhesion with other materials are needed. Other propylene-based polymers, such as ethylene propylene diene monomer (EPDM) rubber, can also be produced by catalyst design of Z-N catalysts or post-metallocene catalysts [19].

Combined with the PP catalyst technology, a series of PP polymerization processes have been developed and commercialized successfully. And the polymerization processes are highly dependent on the PP catalyst system. Typically, in a gas phase, slurry processes, such as Spheripol (Basell), Hypol (Mitsui Chemicals), Unipol (Dow Chemical), Innovene (INEOS), Novelen (BASF), Spherizone (Basell), and Borstar (Borealis), are required to meet the requirements for industrial equipment, such as shape control of products, avoiding reactor fouling with low investment and operating costs and without environmental impact, etc., and heterogenization of Z-N or metallocene catalysts. Isotacticity polypropylene (iPP), random copolypropylene (RPP), impact PP can be produced in these processes, while in contrast, molecular catalysts such as CGC and post-metallocene catalysts are directly used in a solution polymerization process.

2. General structure and properties

PP is a semi-crystalline thermoplastic resin with a linear chain structure consisting of C and H elements. Similar to PE, PP has great chemical resistance toward solvents, acid, and alkali. The alignment of methyl groups attached on the chain backbone, however, may greatly influence the polymer's properties in several ways, including the introduction of a steric group and different stereostructures as shown in **Figure 3**. There are three main different stereoisomers of PP, isotactic PP (iPP), syndiotactic PP (sPP), and atactic PP (aPP) [16]. For iPP, all methyl groups are arranged on the same side of the polymer backbone, in sPP the methyl groups are located on alternating sides, while in aPP, the methyl groups are scattered randomly along the polymer chain. Compared to PE, iPP has higher melting point and modulus due to their stiffening chain and the helical crystal structure. In addition, chain scission rather than cross-linking happens in thermal and high-energy treatment due to the tertiary C atom.

Stereoregularity of the methyl group branch separates crystallizable subspecies from amorphous subspecies; the melting point and modulus strength of sPP is lower than iPP, and among them, aPP has the lowest melting point.

Similarly, copolymerization of propylene and α-olefin with various compositions can vary the crystallizing ability of the polymer chain. There are two general types of polypropylene copolymers: random copolymers and block copolymers. The

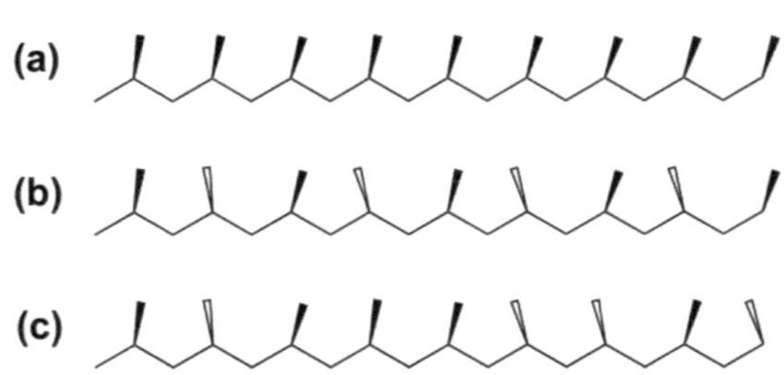

Figure 3.
Three types of stereoisomers of general PP chain: (a) isotactic, (b) syndiotactic, and (c) atactic [16].

comonomer used is typically ethylene, and in the random copolymers, the ethylene content is usually less than 7% [20]. Randomly polymerized ethylene monomer added to polypropylene homopolymer decreases the polymer crystallinity and makes the polymer more transparent. The impact copolymers of propylene-ethylene are virtually a blend of block EP rubber, HPP, and random copolymer, based on the granule reactor technology, so the impact copolymers can obtain great impact strength due to EP rubber phase spread in the PP matrix. A third comonomer can also be used as in ethylene-propylene rubber or EPDM increasing its low temperature impact strength.

3. PP catalyst development and catalyst-defined polymer design

3.1 Z-N catalysts

The massive production of PP is predominated by the use of $MgCl_2$-supported Z-N catalyst systems. These systems consist of a supported catalyst composition, formed by reaction of a transition metal halide (usually $TiCl_4$) and an internal electron donor (ID), generally a Lewis base as a support (typically $MgCl_2$), an alkylaluminum cocatalyst (e.g., triethylaluminum), and an external donor (ED), generally another Lewis base (e.g., alkoxysilanes). These catalyst compositions are independently added in the process of polymerization. A timeline overview on the historical progress in the commercial research and development of Z-N catalyst systems is presented in **Table 1**.

The first major improvement in the Z-N catalysts, which is based on research by Montedison (now LyondellBasell) in Italy and Mitsui in Japan, occurred in 1968 with the discovery of the milled $MgCl_2$ support for ethylene polymerization. This technology was extended to the PP industry in 1970s by extra addition of internal and external electron donors to ameliorate the isotacticity of PP without inhibiting catalyst activity, which led to the third-generation of PP catalysts with high yield (15–30 kg PP/g cat), eliminating the need for catalyst residue removal (Ti $\leq$ 5 ppm), but the atactic component was still inconveniently too high.

The second breakthrough came out with "the reactor granule technology" (RGT) in the 4th generation catalysts in the 1980s. This heterogeneous catalysis was based on active $MgCl_2$ and allowed for a real process simplification, eliminating the process for ash content and atactic removal and avoiding the occurrence of a large amount of fine powder. The particle morphology of the prepared catalyst could be replicated in the final polymer as the particles grow during the polymerization, which was called "replication phenomenon." The catalyst can have a granular or spherical form with a higher and longer activity (20–60 kg.PP/g.cat); also, it has high stereoregularity with isotactic index typically above 95%, tunable molecular weight distribution, and so on [21].

Several routes have been employed for the preparation of granular or spherical catalysts with controllable particle size and morphology. One feasible strategy is implemented by controlled preparation of a new support material, the adducts of $MgCl_2$ and an alcohol, which is subsequently titanated and reacted with an internal donor to obtain supported catalyst. The archetypal example of controlled precipitation was disclosed in the work of Kashiwa and coworkers at Mitsui [22]. $MgCl_2$ is contacted with 2-ethylhexanol in alkane solvent, forming a homogeneous solution. The formed solution then is reacted with phthalic anhydride and $TiCl_4$. The mixture is subsequently contacted with diisobutyl phthalate (DIBP) as an internal donor to form precipitated solid particles with heating. Then the particles are treated

Generation	Composition and structure	Productivity[a] (kg PP/g cat)	II (wt%)	Mw/Mn	H_2 response	Technology control	Process requirements
1st (1957–1970)	δ-$TiCl_3$.0.33$AlCl_3$ + $AlEt_2Cl$	0.8–1.2	88–91		Low	Irregular powder	Need of purification and atactic removal
2nd (1970–1978)	δ-$TiCl_3$ + $AlEt_2Cl$	10–15	94–97		Low	Irregular powder	Need of purification and atactic removal
3rd (1978–1980)	$TiCl_4$/benzoate/$MgCl_2$ + $AlEt_3$ + benzoate	15–30	90–95	8–10	Low	Regular/irregular powder	No purification, need of atactic removal
4th (1980) RGT	$MgCl_2$/$TiCl_4$/phthalate + $AlEt_3$/silane three-dimensional catalyst granule architecture (RGT)	20–60	95–99	6–8	Medium	Particles with regular shapes and adjustable size and PSD. Designed distribution of the different products inside each particles	No purification, no atactic removal
5th (1988) RGT	$MgCl_2$/$TiCl_4$/diester + $AlEt_3$ $MgCl_2$/$TiCl_4$/diester + $AlEt_3$/silane	50–130	95–99	4–6	Very high		
(1999) RGT	$MgCl_2$/$TiCl_4$/succinate + $AlEt_3$/silane	40–70	95–99	10–15	Medium		
6th Phthalate replacement	$MgCl_2$/$TiCl_4$/phthalate-free donor + $AlEt_3$/silane				Very high	Designed distribution of the different products inside each particles	No purification, no atactic removal

[a]*Polymerization conditions: liquid propene, 70°C, 2 h.*

Table 1.
Performance development of Ziegler-Natta catalysts for polypropylene.

with $TiCl_4$ again to obtain the final catalyst. Another successful example using this precipitation method is the N series of catalysts (BRICI, Sinopec). According to this method, anhydrous $MgCl_2$ is reacted with tributyl phosphate and epichlorohydrin in toluene to form a uniform solution. The solution is subsequently treated with phthalic anhydride and $TiCl_4$. The resultant solid particles have regular spherical form; then the solid catalyst is contacted with DIBP and $TiCl_4$ to obtain the final catalyst [23].

The spherical $MgCl_2$•nEtOH support is also a commercially successful example by Basell/Avant ZN range, Sinopec/BRICI DQ catalyst [24], and Brorealis [25]. Spray-drying or controlled precipitation and emulsion process can be adopted to produce the spherical $MgCl_2$•nEtOH support. As exemplified in Avant ZN range, molten emulsions of the $MgCl_2$•nEtOH ($n \approx 2$–3) adduct in paraffin oil are cooled rapidly to obtain spherical particles with a narrow particle size distribution. Similarly, the spherical support of $MgCl_2$•nEtOH ($n \approx 3$) is used in the DQ catalyst [26–28]. Instead of contacting a solid support with catalyst components, the emulsion-based catalyst of Borealis is based on a liquid/liquid phase system, and the catalyst components are contained in a liquid phase.

Furthermore, $Mg(OEt)_2$ has been successfully used as a starting material in SHAC and Toho Titanium THC catalyst system [29–35]. $Mg(OEt)_2$ support plays a significant role in the development of the super high activity catalyst (SHAC) system, with ethylbenzoate as the internal electron donor in the early stage. The internal electron donor ethylbenzoate can be replaced easily by other donors; for example, benzoate is used in commercial SHAC 310 catalyst and phthalate in SHAC 320 catalyst. As a starting material, $Mg(OEt)_2$ also could be converted to a carboxylate by contacting with CO_2, to form a soluble Mg medium. The soluble carboxylate Mg can be reacted directly with $TiCl_4$, which shapes the basis of the Amoco CD catalyst [36–38]. After reacting with a Grignard solution such as nBuMgCl, the $Mg(OEt)_2$ could be converted to a Mg(OR)Cl support with controlled morphology. A catalyst based on this Mg(OEt)Cl support has also been mentioned by SABIC [39, 40] and Basell/Akzo Nobel [40, 41].

The 4th RGT catalysts have also promoted the revolutionary development of PP-based production processes, such as Hypol, Unipol, Spheripol, Novelen, Spherizone, Catalloy, etc., and made it possible to generate multiphase alloys and blends directly in reactors, producing high-performance materials not available from conventional technologies. First, the catalyst is the only active center, propylene polymerization takes place in the catalyst, as the polymerization goes on, the catalyst grows into a polymer particle with active site within it, so both the catalyst and the polymer particle can act as reactor during the polymerization.

For Z-N catalysts, the internal and external electron donors are very critical in tuning the chain structure, MWD of the polymer, and hydrogen response; their effect is mainly determined by the binding energy and mobility on the $MgCl_2$ surface, controlling the stereoregularity of the PP chain. As mentioned above, internal electron donors are added during catalyst preparation while external electron donors are used in the process of polymerization. For a given system, changing the internal donor/Ti ratio and/or Al/external electron donor might lead to a dramatic difference in the polymer structure.

The search for further catalyst improvements brought to light various novel internal donors which were not readily extracted from the support by the alkylaluminum cocatalyst. The diether compounds used as an internal donor in the 5th generation Z-N catalyst systems, in particular 2,2-disubstituted-1,3-dimethoxypropanes with an O–O distance in the range of 2.8–3.2Å, which are not extracted easily due to the relatively strong interaction with the catalyst surface, when contacting the $AlEt_3$ cocatalyst, show high stereospecificity even without using an external donor [42–47]. The diether compounds exhibit particularly high polymerization activity, typically giving yields exceeding 100 kg PP/g cat, good high hydrogen sensitivity, and relatively narrow MWD. Aliphatic dicarboxylic ester-based internal donor and, in particular, succinates and polyol esters have been employed by Basell [48, 49] and BRICI/Sinopec. Different from the 4th generation phthalate-based catalysts, the succinate internal donors developed by Basell produce PP with much broader MWD 10–15 using an alkoxysilane as an external donor. In addition, the succinate-based catalysts also generate ethylene-propylene copolymers with lower glass transition temperature, which enables the production of heterophasic copolymers having great balance of stiffness and toughness. Similarly, the polyol ester family developed by BRICI/Sinopec as an internal donor has a similar polymerization performance to succinate-based systems. However, unlike the succinates, this catalyst system yields high stereoselectivity even without alkoxysilane external donors [50]. In addition to this, mixed donors are also employed in many cases, for example, the mixture donor system of succinate and diether [51] or succinate and dimethoxytoluene [52].

In recent decades, numerous researches focus on finding more potent electron donors for the 6th generation Z-N PP catalysts. Despite the fact that the phthalate-based catalysts produce PP with far lower phthalate content below the 0.3 wt% (3000 mg/kg) concentration limit in the REACH Regulation (EC) 1907/2006, a totally phthalate-free solution is highly motivated and becomes a competitive advantage. As shown in **Figure 4**, there are several types of nonphthalate electron donors that are commercially used or proposed [53]. The new nonphthalate solution (Consista donors) of Dow (Now Grace), as a exemplified example 1,2-phenylene dibenzoate donors (**Figure 5**) [54–57], undoubtedly takes the vanguard in the developments of nonphthalate replacement. It is worth noting that a more complicated catalyst preparation process may be required in contact with the internal donors and $TiCl_4$. In order to improve the final polymer particle morphology and catalyst performance, the process of use of ethylbenzoate or 2-methoxy ethylbenzoate as buffering to the procatalyst and $TiCl_4$ may be needed before adding the phenylene dibenzoate [58].

Typically, alkoxysilane compounds are used as the external donors and added in the polymerization process. The stereospecificity of the Z-N catalysts could be controllable by changing the substituents of the alkoxysilanes containing relatively bulky groups [53]. The correlations between the structure of the silanes' external donor and their polymerization performance have been discussed in detail by Härkönen and Seppala [59–65]. Silanes containing hydrocarbon substituents and oxygen atom with the appropriate size and electron density are expected to obtain PP with high isotacticity index. An industrial silane external donor typically contains at least one secondary or tertiary carbon linked to the silicon atom. It is reported that this bulky group could protect the silane against removal from the catalyst surface when contacting with aluminum alkyl [66]. By far, cyclohexyl(methyl)dimethoxysilane (donor C) and dicyclopentyldimethoxysilane (donor D) have been most commonly used [66]. Some commercial external donors are listed in **Figure 6**. When compared to donor D of the polymerization performance, donor C gives high hydrogen sensitivity and the latter gives particularly high stereospecificity [67] and broader MWD [68].

The Mw and MWD of a PP are critical to the end-use application of the PP product. For fiber spinning applications, relatively low MW and narrow MWD are favored. In contrast, high melt-strength is required for extrusion of pipes and thick sheets; therefore, broad MWD and relatively high MW are needed. For heat-resistant PP, generally with high isotactic stereoregularity, broadening MWD is beneficial for balance between high rigidity and toughness. And choosing

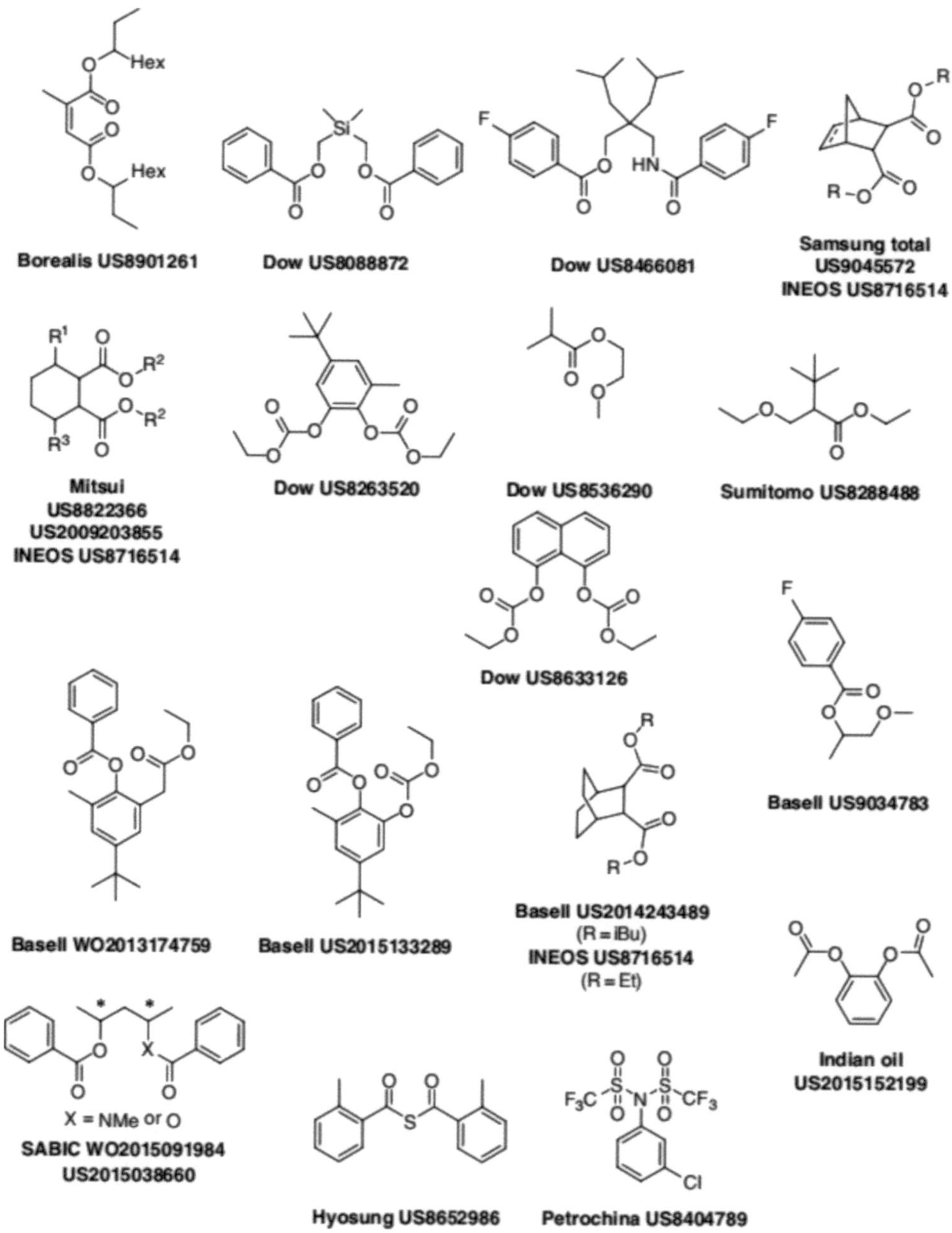

Figure 4.
Recent disclosure of internal donors [53].

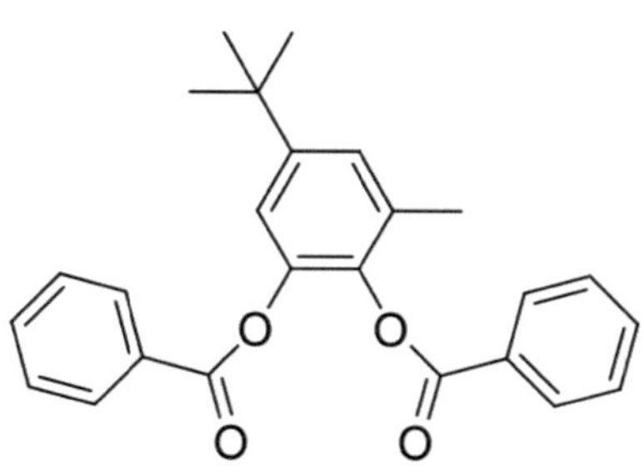

Figure 5.
1,2-Phenylene dibenzoate donors from Dow.

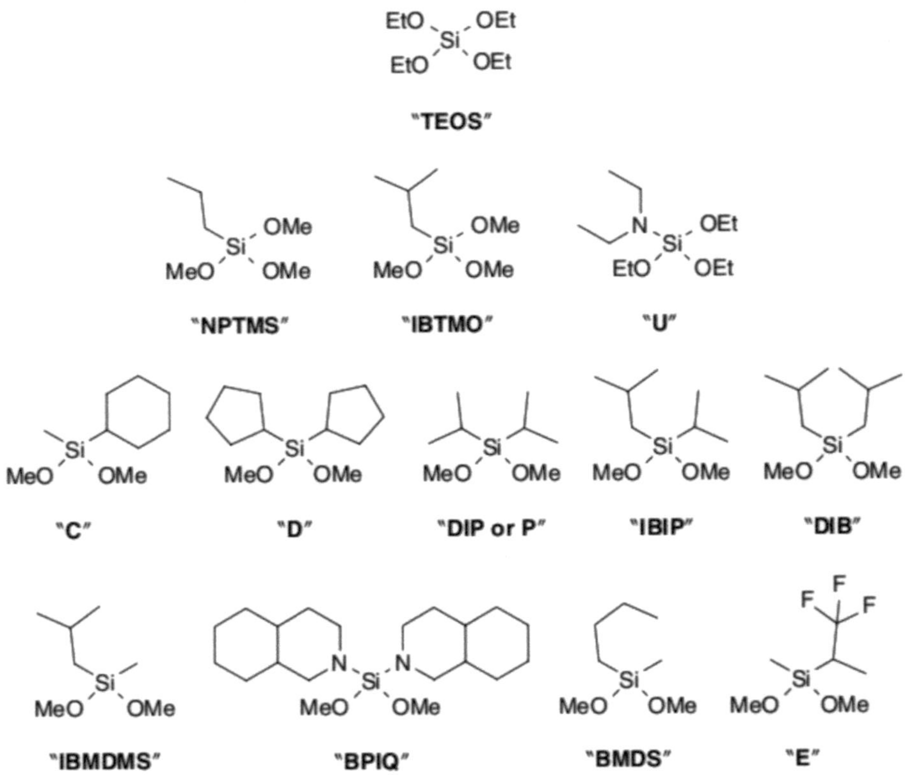

Figure 6.
Typical industrial external donors in use.

alkoxysilane external donor in the polymerization process is a convenient approach to control stereoregularity, MWD, and the H_2 response and ethylene incorporation. However, it is often the fact that a single external donor is difficult to yield desirable control over multiple properties of the final polymer. To overcome this problem, multiple external donors could be taken into consideration to tune polymer properties by mixed external donors or separate addition in different reactors. Take for example that Exxon have exploited combination of tetraethoxysilane (TEOS) and donor D in a two-stage polymerization process [69]. TEOS alone was utilized at the first stage, producing high melting index polymer; then, donor D was added at the second stage, forming resin with high isotacticity and low melt flow rate, similar to those produced by D donor alone. However, the combination of multiple donors and a two-stage process obtained a final polymer with high isotacticity, high melting index, and broader MWD. In addition, the mixed donor D and TEOS systems showed higher incorporation of ethylene and high hydrogen sensitivity in continuous copolymerization of propylene with ethylene [70]. Another case in point of combination of multiple external donors was revealed in the "self-extinguishing" catalysts concept of Dow's SHAC™ catalyst system in the Unipol process [71]. In the process of polymerization, a mixed external donor consisting of alkyl benzoates and alkoxysilanes is used; the catalyst systems are very active at the normal operating conditions but dramatically lose their activity at higher temperatures, and therefore prevent reactor-fouling. Further development in the "self-extinguishing" catalysts is also mentioned by combinations of NPTMS with aliphatic esters such as di-n-butyl sebacate or isopropyl myristate [72].

3.2 Metallocene catalysts

Since the discovery of MAO by Sinn and Kaminsky, single site catalysts, and metallocenes particularly, have presented significant and meaningful innovation in olefin polymerization catalysis. For commercialization of metallocene catalysts, these metallocene systems have to be compatible with the advanced process technologies, which are referred to as "drop-in catalysts."

From a commercial perspective, iPP seems a reasonable starting point, and metallocenes suitable for iPP production generally are based on supported-zirconocenes. So far, over 900 applications since 1984 have been patented on the iPP-based metallocenes by Hoechst, Exxon, Fina, Mitsui, BASF, and so on. The typical chemical structure of the employed zirconocenes is illustrated in **Figure 7**.

The single-site metallocenes allow microstructure tailoring in the molecular level of the produced PP with narrow molecular weight distribution (Mw/Mn = 2). The chance to control the polymer molecular structure by metallocene complex design helped to shape a better knowledge of the basic structure-performance relations in PP. As illustrated in **Figure 8**, different molecular chains of PP with various stereoregularity can be produced by design of the metallocene catalyst structure [73–76], such as isotactic polypropylene (iPP) [77, 78], hemi-isotactic polypropylene (hiPP) [79], syndiotactic polypropylene (sPP) [80], and stereoblock polypropylene [81–83]. The correlation between the microstructure of PP and the symmetry of the metallocene complexes has been established via Ewen's stereocontrol rules [84]. These tailor-made metallocene catalysts indicated the 2nd breakthrough in olefin polymerization followed by the discovery of MAO.

Along with the development of the tailor-made metallocenes, the homogeneous stereospecific catalysis has allowed the industrial preparation of atactic polypropylene elastomers with high molecular weight and longer iPP blocks. Several metallocene complexes have been employed for producing PP-based elastomer by designing the architectures of the metallocene catalyst. A "molecular switch" approach was reported by Waymouth and Coates, the unbridged metallocene catalyst can be changed into a dual-site catalyst by intramolecular oscillation to generate stereoblock PP without extra metallocene [85–89]. A schematic illustration is shown in **Figure 9**, and these complexes were thought to oscillate stereocontrol between their rac-like (isospecific) and meso-like (nonstereospecific) configurations during polymerization, thus producing stereoblock PP containing alternating isotactic and atactic blocks. This intriguing idea was further developed by other groups to discover other metallocenes having similar behavior of oscillating stereocontrol, such

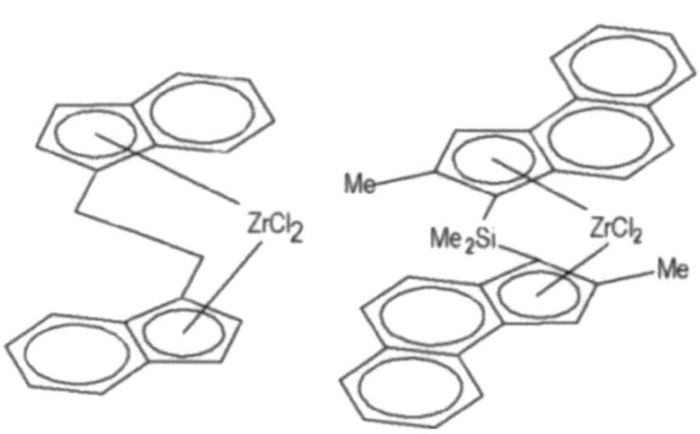

Figure 7.
Hoechst (left) and BASF metallocenes [9].

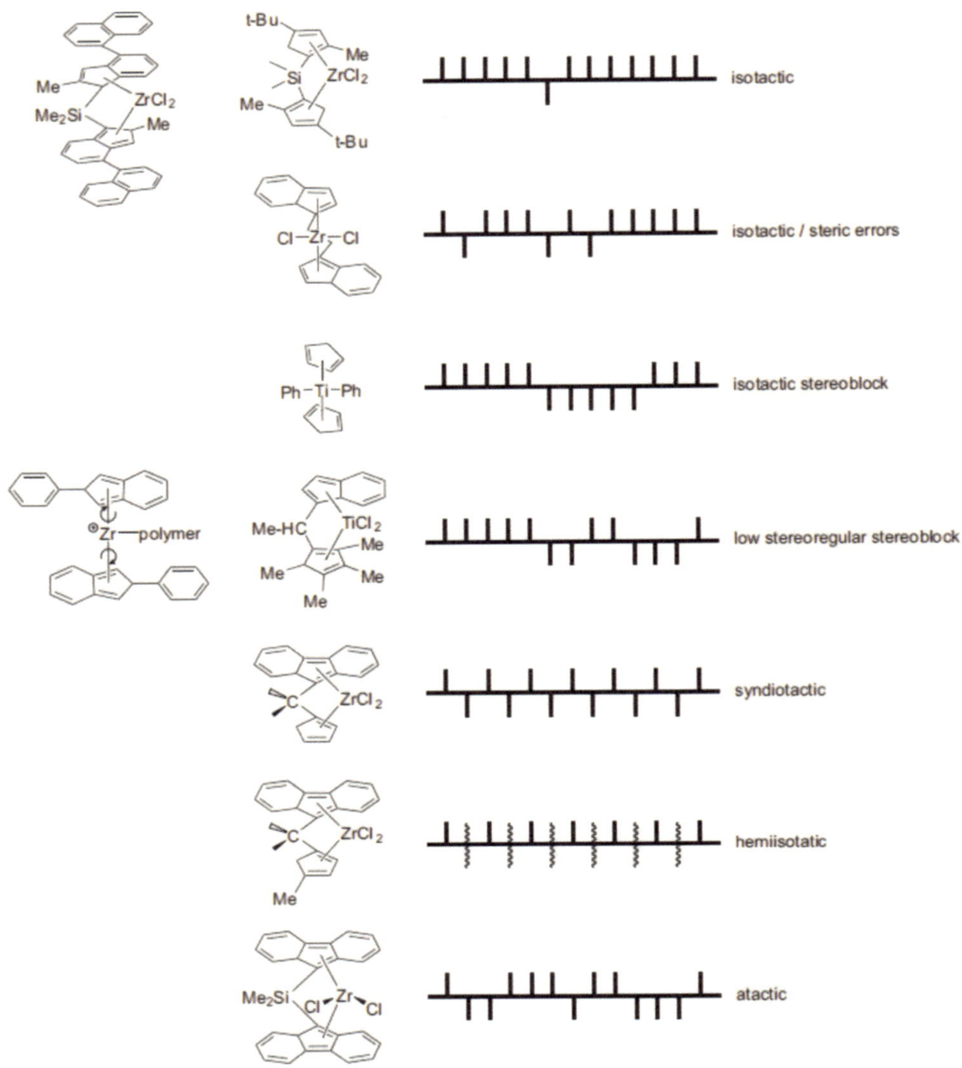

Figure 8.
Correlation between metallocene structures and polypropylene architectures.

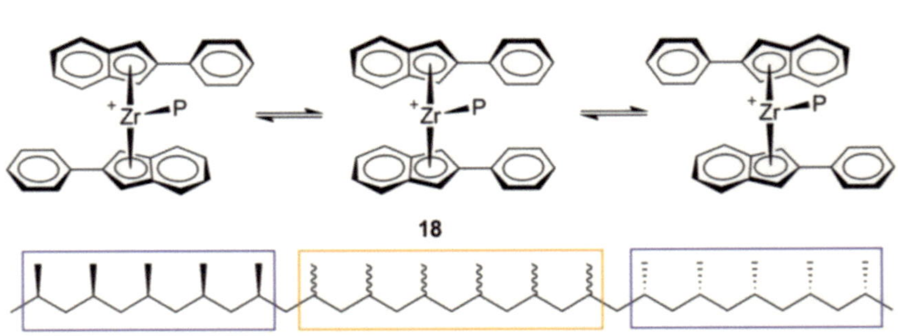

Figure 9.
Molecular switching between the two site configurations during polymerization of the unbridged metallocene catalyst.

as bis(2-aryl-indenyl) zirconocenes and hafnocenes with diverse aryl substitutions [90–92]. Moreover, the types of solvents and counterions can unfavorably limit ligand rotation; therefore, PP with complex microstructures might be produced.

By combining different stereospecific metallocenes, thermoplastic PP elastomers could also be prepared. In 1996, Chien employed a mixture of two different metallocenes, which consists of isospecific C_{2v}-symmetric metallocenes such as rac-ethylenebis(1-indenyl) zirconium dichloride or rac-dimethyl silylbis(1-indenyl) zirconium dichloride and nonstereospecific C_2-symmetric metallocene such as ethylenebis(9-fluorenyl)zirconium dichloride, to produce PP-based elastomers by propylene homogeneous polymerization [93–95]. The combined dual-site metallocene catalysts generate a reactor blend comprising isotactic PP, atactic PP, and stereoblock PP containing alternating isotactic and atactic blocks. In the produced blends, the stereoblock PP contains immiscible isotactic and atactic PP blocks due to the chain transfer between stereospecific and nonstereospecific sites. Different from the early multisite catalysts for generating PP-based elastomer, this dual-site catalyst system, which combines two single-site metallocene catalysts, allows flexible tailoring of the elastomer properties by simply changing the molar ratio of added C_2/C_{2v} metallocenes. Surprisingly, Fink successfully synthesized iPP-b-sPP stereoblock PP using silica-supported MAO/dual-site metallocene catalysts that combined isospecific rac-$Me_2Si[Ind]_2ZrCl_2$ with syndiospecific iPr(FluCp)$ZrCl_2$. In another example of producing iPP-b-sPP stereoblock PP, Rytter and coworkers used $AlMe_3$ together with the MAO as an activator to the homogeneous dual-site catalyst system consisting of isospecific rac-Me_2Si(4-t-Bu-2-MeCp)$_2ZrCl_2$ and syndiospecific Ph_2C(FluCp)$ZrCl_2$ [96].

Many novel polymers can be produced by post-modification reactivity of metallocene-generated polymers. For example, vinyl-terminated PE could be incorporated onto a polypropylene backbone by a metallocene catalyst [97]. However, due to poisoning effects, polar groups should be avoided in the process of olefin copolymerizations, but protecting routes such as polar grafted PP can be employed in their incorporation [98]. Furthermore, 1,4-divinylbenzene comonomer is often used in propylene polymerization to generate styryl-capped chains that can be further functionalized at the chain ends [99]. Chain ends by chain transferring to alkyl aluminum can also be functionalized by a variety of reactions [100].

3.3 Post-metallocene catalyst

Like the single-site metallocene, the post-metallocene catalysts also possess the ability to tune the structure and properties of the prepared PP with precision, through tailoring the substitution patterns of the ligand [101–103]. Syndiotactic polypropylene (sPP) was produced by a phenoxy-imine ligand system discovered by scientists at Mitsui and simultaneously at Cornell University in New York. Subsequently, a series of researches and development focusing on post-metallocenes resulted in highly isotactic PP with high activity (**Figure 10**) [104–106].

In 2004, based on the Symyx Technologies (Symyx) using high throughput technology by Dow, VERSIFY™ Plastomers and Elastomers which are based on propylene/ethylene copolymers were developed with a pyridyl amine catalyst system (**Figure 13**, the 2nd complex). The produced VERSIFY™ Plastomers and Elastomers possess narrow molecular weight distribution, a broad chemical composition distribution compared with polymers from other single-site catalysts, and unique regio defects, which are very critical to their properties. These novel materials are the first PP-based polyolefins made in Dow's solution process, and they have excellent optical properties with high clarity and great processing performance compared to their ethylene-based counterparts. Although high temperature performance, elastic recovery, and anti-scratching and mar resistance performance are inadequate in many applications in high value fields due to their high ethylene-propylene rubber phase, they have been expanded to many new applications as new materials owing to their brilliant toughness, the optical properties, and filler compatibility.

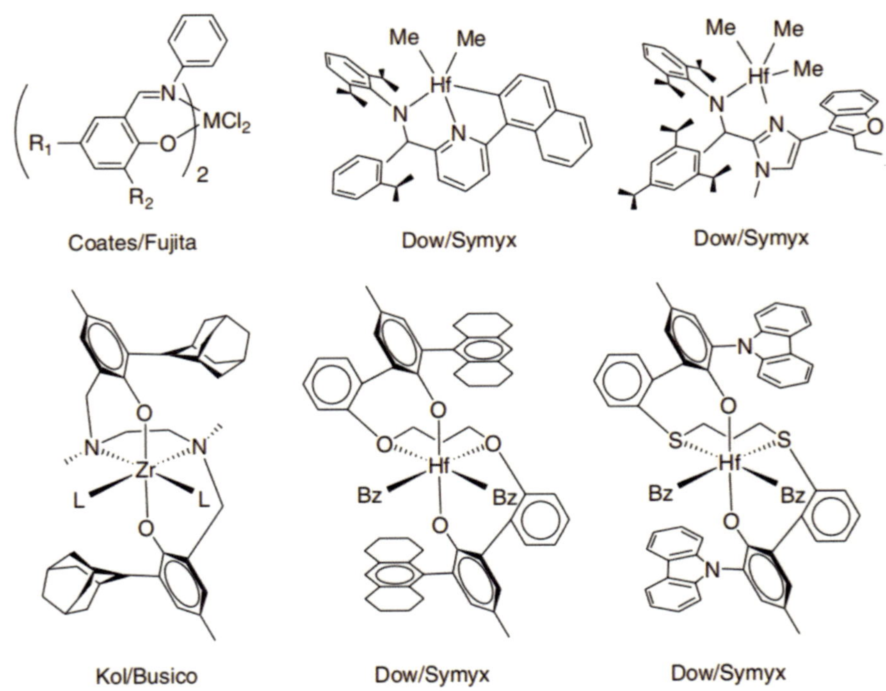

Figure 10.
Post-metallocenes for isotactic PP.

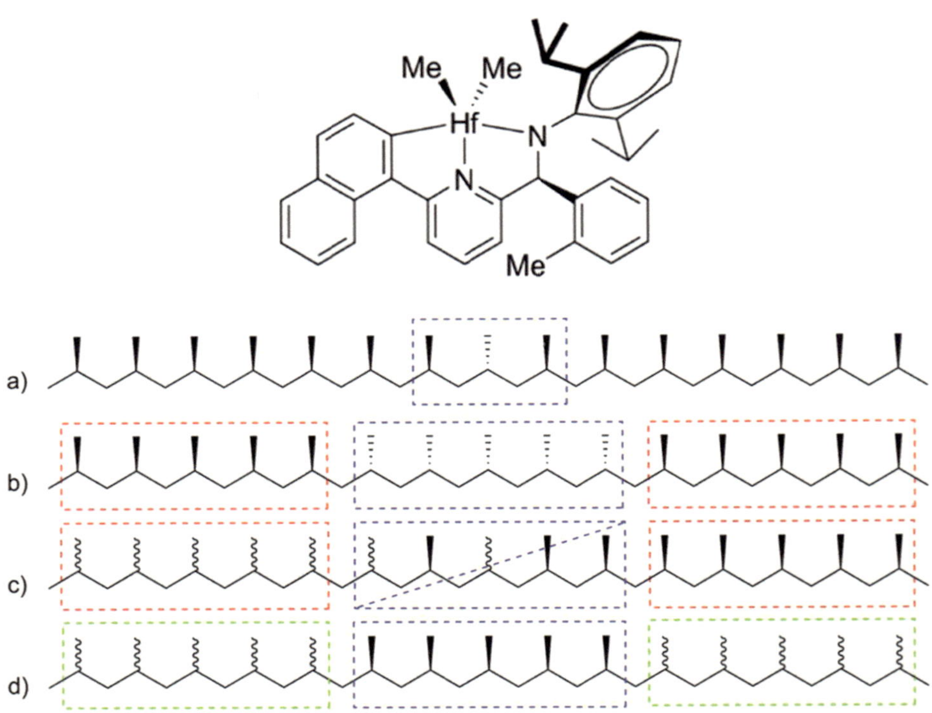

Figure 11.
Polypropylene with different stereoisomers: isotactic polypropylene with isolated stereodefects (a), isotactic stereoblock polypropylene (b), gradient (c), and stereoblock polypropylenes (d), which contain hard isotactic and flexible atactic segments [3].

Inspired by the chain shuttling technology, Dow then using the pyridyl amine catalyst system combined with alkylaluminum as chain shuttling agent provided an alternative way to make stereo-block polypropylene with narrow MWD and high molecular weight [108–110]. As illustrated in **Figure 11**, single-site racemic catalysts are used to prepare stereoblock PP, and the catalyst system actually is a dual-site catalyst containing a 50/50 mixture of the two enantiopure sites, yielding isotactic stereoblock PP. In the solution polymerization, polar solvents such as 1,2-difluorobenzene are preferred, in order to improve alkylaluminum-mediated polymeryl shuttling between the two sites, giving rise to alternating isotactic blocks with either (R) or (S) configuration [111, 112]. There are also other combined pyridyl amine catalysts with CGC or metallocene catalysts employed in the chain shuttling polymerization, to produce stereo-block polypropylene [3].

Ketimide and amidinate complexes can also be used for ethylene propylene diene monomer (EPDM) in a high temperature solution polymerization process, and the highest activities in ethylene polymerization were achieved with bis(tert-butyl)ketimide ($N{=}C\text{-}t\text{-}Bu_2$)-ligated titanium complexes [19]. This type of ketimide catalysts is licensed from Nova Chemicals by DSM Elastomers for the production of a new EPDM rubber product, with the trademark of Keltan ACE (referred to as "advanced catalysis elastomers"). The catalyst structure was further modified to amidinate complex 9 with the general structure $[(C_5R_5)Ti\text{—}\{N{=}C(Ar)NR'_2\}X_2]$ (X = Me or Cl) [113–116].

4. PP polymerization process

The production of PP and PP-based polymers in commercial scale is highly related to catalyst and process technology. Typically the processes could be divided into four categories: gas-phase, bulk, slurry, and solution polymerization technologies.

For homopolymer and random copolymers, bulk, slurry, and gas-phase processes can be employed. By connecting an additional gas-phase reactor to the polymerization equipment, in which the EPR is generated by ethylene-propylene copolymerization, impact PP can be produced. For the production of a polypropylene-based elastomer and most of the commercial EPDM rubbers, a solution process is required with a homogeneous catalyst system. In a slurry, bulk, or a gas-phase reactor, the polymer is generated around the heterogeneous catalyst particles. The slurry process and bulk process typically employ autoclaves or loop reactors. Gas-phase reactors generally adopt the form of fluidized-bed or stirred-bed. In fluidized-bed reactors, a gaseous stream of nitrogen and monomer is responsible for fluidizing the polymer particles and transferring the reaction heat, while in a stirred-bed reactor, mechanical stirring is employed to distribute the polymer particles and transfer heat. In a specific gas-phase stirred-bed reactor, horizontal or vertical layout can be taken. Different reactor configurations are illustrated in **Figure 12**.

The polymerization can be conducted in a single reactor or multiple reactors. Single reactors are typically employed to produce uniform composition, while multiple reactors in series can be adopted for PP with more complex microstructure and composition distribution expanding the properties of homopolymers and copolymers. Impact PP copolymers are typical examples produced in multiple reactors. The target product can be achieved in the two different steps diversifying the polymer microstructures. Isotactic polypropylene particles are formed in the first reactor by propylene homopolymerization, while a stream of mixture of propylene/ethylene is fed to the second reactor to make propylene-ethylene rubber phase copolymers, which is dispersed within the same catalyst and homopolymer particles. A schematic illustration of impact PP production is shown in **Figure 13**.

The improved impact strength attributes to the heterophasic ethylene-propylene copolymer scattered in the crystalline iPP matrix formed in the first reactor.

In the past decades, the process configuration has evolved, driven not only by economics, but mainly by product performance. Multiple reactors allow for flexible operation during polymerization to make various polymers that can be considered a blend of multiple components (reactor blends). Along with the improvement of the catalyst technology, the major worldwide companies have developed their own PP process using multiple-stage technology or combining bulk and gas-phase polymerization: Spheripol (Basell), Hypol (Mitsui Chemicals), Unipol (Dow Chemical), Innovene (INEOS), Novelen (BASF), Spherizone (Basell), and Borstar (Borealis).

4.1 Spheripol process

A typical Spheripol II process, which is one of the most employed process owned by Basell, consists of two liquid loop reactors and a gas phase reactor for

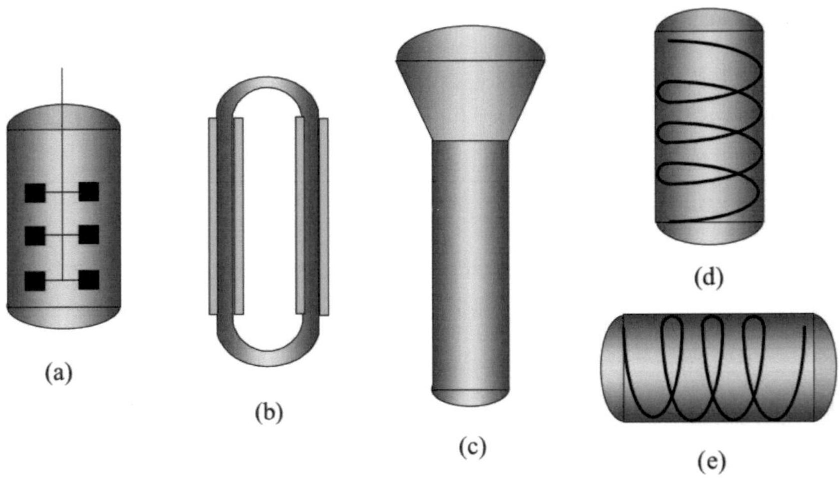

Figure 12.
Reactor configurations for olefin polymerization: (a) autoclave; (b) loop; (c) fluidized-bed; (d) vertical gas-phase; and (e) horizontal gas-phase [107].

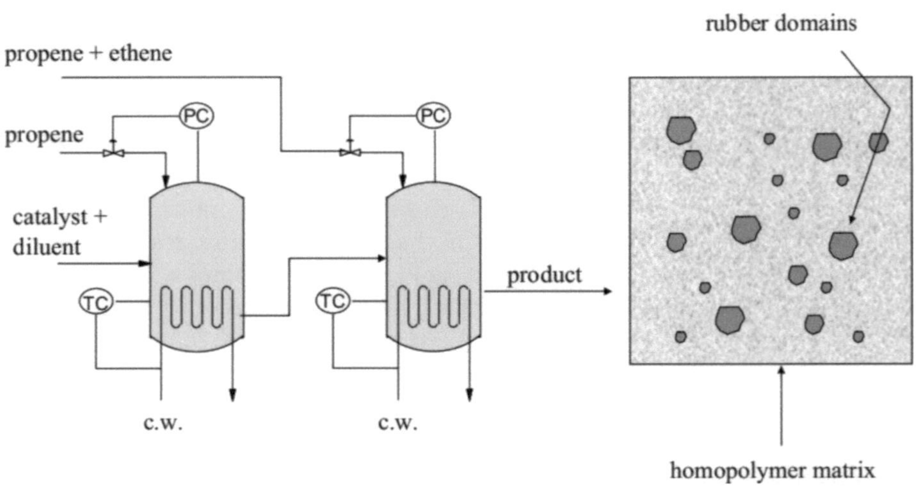

Figure 13.
Process of production of impact PP [107].

producing more than 100 types of PP brands including homopolymer and random and impact copolymers with supported catalysts. The two loop reactors are typically operated under standard conditions of 70°C and 4 MPa with liquid propene circulating round the two loop reactors. An axial flow agitator is used to maintain good flowability, ensure good heat transfer, and keep substantial polymer particles from agglomerating from the slurry in each loop [35]. Typically, the PP concentration content was kept around 40 wt%.

HY/HS catalysts, together with triethylaluminum, and an external donor such as a dialkyldimethoxysilane are continuously fed into the reactor to produce PP with a specific stereoregularity. Morphology control of the HY/HS catalyst particles is very critical in the initial polymerization stage (the first several seconds to minutes). For this reason, a prepolymerization stage in which the catalyst particles react at lower temperature and monomer concentration is employed to prevent catalyst particles from generating fine powders due to rapid polymerization. The catalyst particles grow slowly in this prepolymerization stage and produce only small amounts of polymer (< 100 g/g) in the catalyst granules before this catalyst is fed into the loop reactor. Mean residence time of the polymer particles in a single loop is about 1–2 h. Multiple loop reactors can be operated in a tandem way to narrow residence time distributions, tune the polymer structure and properties, and increase productivity.

The isotactic PP particles are produced in the two loop reactors bypassing the copolymerization unit. Due to the high isotactic stereoregularity, generally, the HPP obtains high rigidity and tensile strength with high melting point. By adding propylene and ethylene and hydrogen in the two loop reactors, random copolypropylene (RPP) can be prepared; generally, the ethylene content in the RPP is below 4 wt% to avoid reactor fouling. The stereoregularity of a RPP, thus, is impaired via the insertion of the ethylene molecule; therefore, the melting point and crystalline temperature decreased, and the soluble fraction in the temperature-rising elution fraction (TREF) also increased correspondingly. However, the optical properties of the RPP could be improved dramatically with very low gloss, especially when a transparent nucleating agent is added. In the 3rd gas-phase reactor, either a vertical reactor with an agitator or a fluidized-bed reactor, mixture of propylene and ethylene, is fed to generate an ethylene-propylene rubber phase which is dispersed within the isotactic polypropylene particles already formed from the loop reactors; thus, impact polypropylene can be prepared. The EP rubber consists of polyethylene and polypropylene blocks $\{-(CH_2{-}CH_2)_x{-}[CH_2{-}C(CH_3)]_y{-}\}$, and in an industrial production, the ethylene content of the whole impact PP could be up to 15%, or the rubber phase content could be up to 30%.

4.2 Hypol process

Mitsui Petrochemical has developed a similar process using bulk polymerization with their specific supported catalysts. In a Hypol I PP process (**Figure 14**), two conventional reactors in series are employed, with reaction heat removed by evaporation of cool liquid propylene in the reactors. The slurry containing PP particles is then fed into a heated flash vessel, in which the propylene is recovered for recycling use. One or two gas-phase reactors with stirring might be added in tandem with the two bulk polymerization reactors to produce impact polypropylene. The reactors in the 2nd Hypol process of Mitsui Chemicals are replaced with two loop reactors and gas-phase fluidized-bed. Similar with Spheripol process, Hypol processes are design with innovative HY/HS supported catalysts to be able to produce HPP, RPP and impact copolymers.

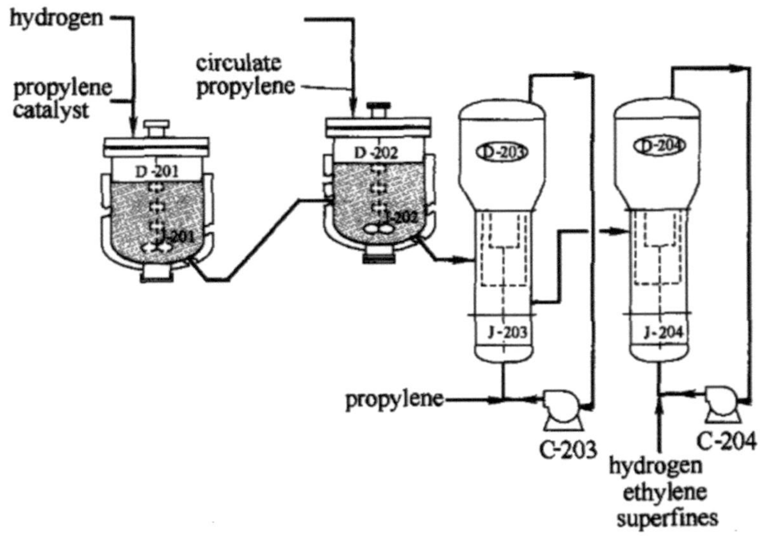

Figure 14.
Hypol I PP process technology.

4.3 Unipol process

Unipol polypropylene process technology, which has its roots in gas-phase polyethylene process of Union Carbide Corporation (UCC), is built around two vertical fluidized bed reactor systems. Dow's Unipol PP process is one of the most advanced processes used worldwide, which allows simplified process scheme and high cost-effective investment with less factory field and equipment. The process is stable and predictable. In the upper section of the vertical gas-phase reactor, there is an expandable pressure vessel, where it operates at 3.5 MPa and 65°C. The heat removal can be improved by charging liquid propylene and propane, and the operating conditions satisfy the dew point of the monomer; therefore, it achieved high energy efficiency. The fluidized-bed reactors have extensive back mixing and a short residence time of less than 1.5 h, which is beneficial to shorten transition product.

With Shell High Activity Catalyst (SHAC) of Dow (Now Grace) for the UNIPOL process, HPP, RPP, and impact PP could be produced using this Unipol technology. Especially, this process is suitable to develop and produce high-performance RPP and impact PP. The ethylene content of a RPP can be up to 7–8%, and be above 20% for an impact PP, and thus, the rubber phase could be above 40% in an industrial production. Impact copolymer is produced in the second reactor in **Figure 15**, where ethylene, hydrogen, and recycled propylene are fed to produce polymer particles containing isotactic PP and active catalysts generated from the first reactor. Then, the reactor polymer product (bottom) is sent to a gas/solid separator where the separated gas is recycled back to the top of the copolymer reactor. The final solid product is discharged from the separator to the purge tower.

4.4 Spherizone process

The concept of multizone circulating reactor (MZCR) was proposed by Basell two decades ago, and the Spherizone process employing the revolutionary reactor aims at improving product properties and broadening the range of products, despite the success of the Spheripol process. The Spherizone process, using Basell's HY and RGT catalysts, involves continuous circulating of the growing reactor granule between two interrelated zones operating under peculiar fluid-dynamic pattern,

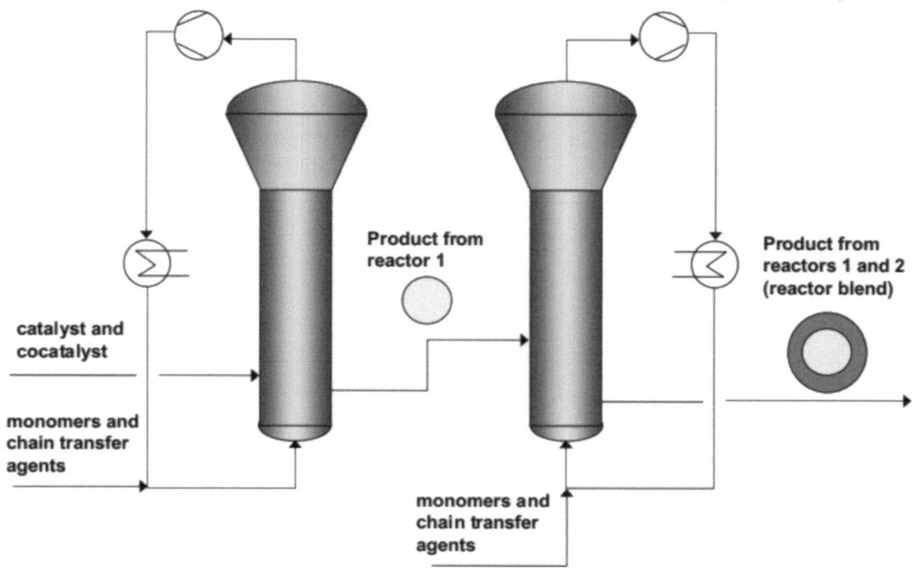

Figure 15.
Unipol gas-phase fluidized-bed process.

and produces polymer particles with excellent morphology and tunable composition in the reactor. The versatility of the Spherizone process is demonstrated by the various high-quality products that include HPP, RPP, and some special products. Due to the unique design and operation, the MZCR process is capable of providing polymers with various components within the same catalyst granule in a short residence time, making them mix at a molecular scale.

In the so-called riser (**Figure 16**), the polymer particles and monomers flow upwards in a fast fluidization state. In the top, the rising gas is separated from the solid particles, which then enters and goes downward in the so-called downcomer. The circulation of the polymer particles is handled by the pressure balance between the two zones. In the multizone circulating reactor, HPP and RPP can be produced. Optionally, the multizone circulating reactor can then be connected with another gas-phase fluidized bed reactor, where ethylene-propylene copolymerization can take place to generate high impact PP. In this reactor, the EP rubber phase is generated within the HPP matrix discharged from the first reaction stage.

4.5 Novelen process

BASF's Novelen PP process is based on two gas-phase reactor vessels equipped with vertical helical agitators, which can provide excellent agitation for catalyst particle dispersion and heat transfer, for making HPP, RPP, and impact copolymers of ethylene-propylene using their specific high activity and stereospecific catalysts. The catalysts need to be well-dispersed in the polymer particle tank to avoid agglomeration of the granule particles, and this process is operated under conditions of 70–80°C and 3–4 MPa to keep the monomer gaseous in the reactors. Reaction heat can be effectively removed from the evaporation of liquid propylene, so the temperature is controlled stably.

The primary reactor can only be used to produce HPP and random copolymers of ethylene-propylene. For impact copolymers, the EP rubber phase is produced in the 2nd vessels, by feeding propylene and ethylene into this reactor.

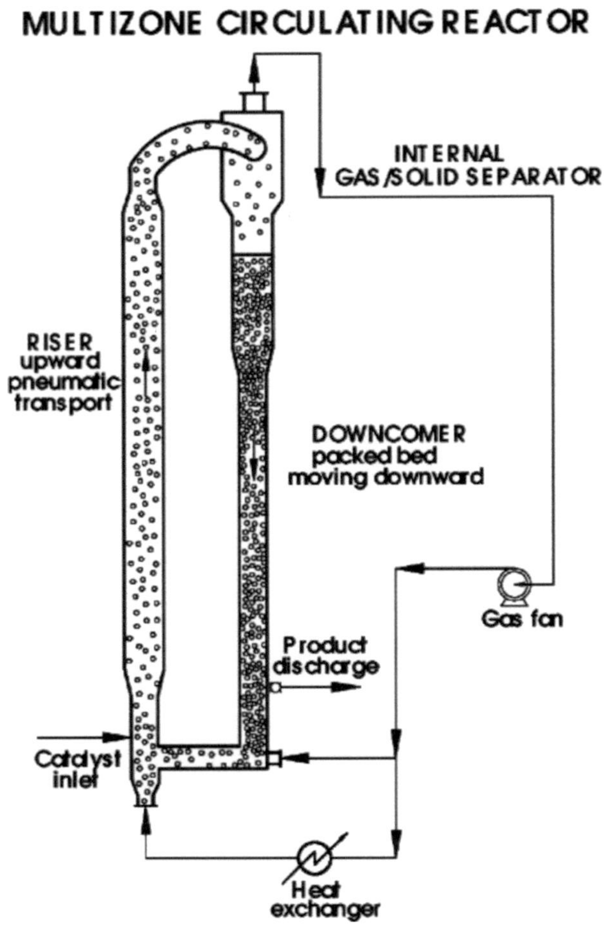

Figure 16.
Multizone circulating reactor [118].

4.6 Innovene process (BP Amoco)

Instead of the vertical agitator of the BASF process, the INEOS's PP process uses one or two horizontally stirred reactors. Uncondensed monomers and hydrogen are injected into the reactors from the base to maintain the gas composition, while condensed recycled monomers are discharged into the top of the reactor to provide cooling. Fluidized-bed deactivation system is used in both horizontal reactors to avoid particle agglomeration. Homopolymers, random copolymers, and impact copolymers can be produced in this process using highly effective CD series catalysts. This process is suitable for the production of high-impact copolymers with high rubber content, and the rubber phase content could be up to 50%.

4.7 Borstar process

The Borstar PP process developed by Borealis is based on a double loop reactor and a fluidized-bed reactor in tandem connection for the production of HPP and RPP. The loop reactor is operated under supercritical conditions with increased comonomer and hydrogen concentration, and higher temperature avoiding gas bubbles. For producing the impact PP, additional one or two fluidized-bed reactors are connected in series, as illustrated in **Figure 17**.

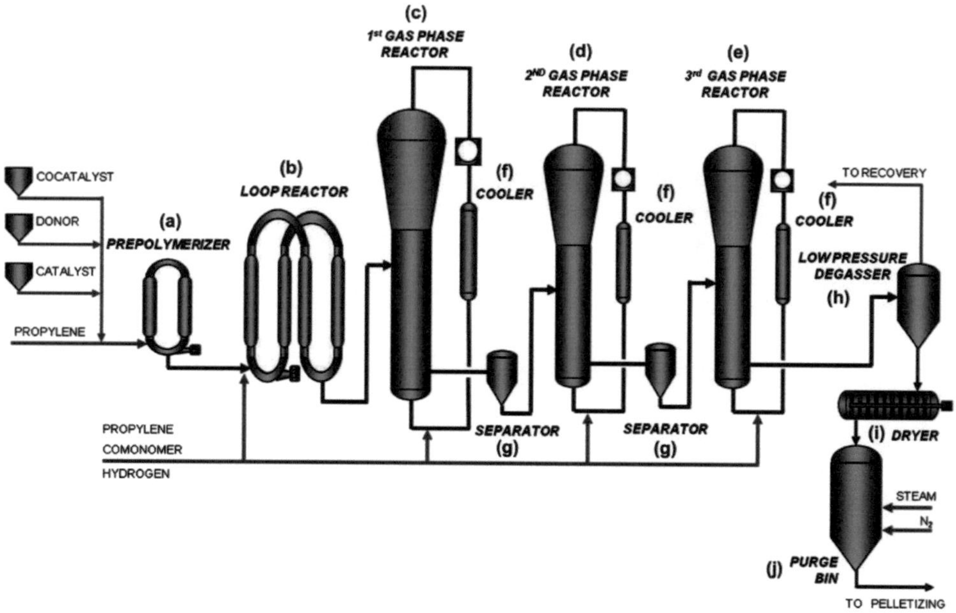

Figure 17.
Borstar polypropylene process scheme (four-reactor setup) :(a) prepolymerizer; (b) loop reactor; (c) first gas-phase reactor; (d) second gas-phase reactor; (e) third gas-phase reactor; (f) coolers; (g) separators; (h) low-pressure degasser; (i) dryer; (j) purge bin [117].

Using the HY/HS catalysts of Borealis, which are prepared from emulsion strategy from two-phase liquid/liquid contact, Borstar PP technology produces high-performance HPP and RPP with very low ash content and narrow molecular weight distribution in the supercritical loop reactor due to excellent heat removal and high productivity. The rubber phase (EPR) of the impact copolymers is typically generated through the 3rd and 4th gas-phase reactors. Using the multi-reactor processes described, it is feasible and practical to further broaden PP material properties meeting up the ever-demanding customer needs.

4.8 Solution polymerization process

Autoclaves, tubular or loop reactors all could be used in a solution process. Different from the slurry and gas-phase process, the solution polymerization process is generally operated at a much higher temperature above 100°C, and in order to form a homogeneous uniform solution, a reaction solvent is added to dissolve the catalyst and polymer generated during the polymerization. Typically, the average residence time is much lower (5–20 min) in contrast with the bulk or gas-phase process (1–4 h). Since polymerization takes place in a more uniform solution, heat and mass transfer resistances are avoided, and this homogeneous process is practical for the production of ethylene-polypropylene copolymer elastomers, EPDM rubbers and stereoblock PP with soluble Ziegler-Natta catalysts, metallocenes, and post-metallocenes. Although conventional HPP could be produced in a solution process, the overall cost is too high to produce them due to requirements for use of solvent and additional process for removal of the solvent.

Solution polymerization process is also very appropriate for production of the atactic PP and PP-based elastomers. Commercial equipment developed to make atactic PP has been announced by Himont (Canada), El Paso (Texas), and Huls (Germany). Special PP-based polyolefins are also produced using the solution process by Exxon, Dow Chemical, and other companies. Propylene-ethylene copolymers referred to as commercialized Vistamaxx™ are launched successfully by ExxonMobil with a special metallocene catalyst. The VERSIFY™ Plastomers and Elastomers of propylene/ethylene-based copolymers are produced profitably with a new post-metallocene pyridyl amine catalyst system in solution process. Stereoblock PPs [17] are also produced by using oscillating metallocene, or a mixed catalyst systems of two metallocenes and/or post-metallocenes in the presence of chain transfer agent or chain shuttling agent, as mentioned above.

5. Conclusion and outlook

Polypropylene-based polyolefins are undoubtedly one of the most robust material fields in polymer production and consumption worldwide. The remarkable progress in olefin polymerization catalyst and PP process technology contributes to the development of various PP-based polyolefins and helps to meet the continuous growing demand of the global market. Today, HY/HS Ziegler-Natta catalysts and RGT technology will continue to play a significant role in production of HPP, RPP, and impact PP especially via bulk, gas-phase polymerization process, or combination of the two processes with the newest developed process technology. With RGT technology which provides virtual nanoreactors within the catalyst and the polyolefin particles, it is possible to design more high performance micro-reactor catalysts. The molecular catalysis of single-site metallocene and post-metallocene catalysts enables precise control of the molecule structure and properties via homogeneous solution polymerization, and high-throughput screening technology provides more robust single-site catalysts with a molecular architecture approach. This allows materials scientists and engineers to design products with a broadening range of properties with advanced catalysts and polymerization process. Despite high activities, a little nontoxic catalyst residue could still be left in the polyolefin; therefore, research and development of greener and cleaner PP products with special high-performance are still way ahead.

Author details

Xiong Wang*, Xiaoyu Han and Renwei Xu
Lanzhou Petrochemical Research Center, Petrochemical Research Institute, PetroChina, Lanzhou, China

*Address all correspondence to: wangxiong1@petrochina.com.cn

References

[1] Sauter DW, Popoff N, Bashir MA, Szeto KC, Gauvin RM, Delevoye L, et al. The design of a bipodalbis(pentafluorophenoxy) aluminate supported on silica as an activator for ethylene polymerization using surface organometallic chemistry. Chemical Communications. 2016;**52**:4776-4779

[2] Wang X, Han X, Ren F, Xu R, Bai Y. Catalysts. 2018;**8**:146

[3] Stürzel M, Mihan S, Mülhaupt R. Chemical Reviews. 2016;**116**:1398-1433

[4] Grein C. Advances in Polymer Science. 2005;**188**:43

[5] Grein C, Gahleitner M. Express Polymer Letters. 2008;**2**:392

[6] Yamaguchi M, Miyata H, Nitta K-H. Journal of Polymer Science. Part B. 1997;**35**:953

[7] Yamaguchi M, Nitta K-H, Miyata H, Masuda T. Journal of Applied Polymer Science. 1997;**63**:467

[8] Phulkerd P, Funahashi Y, Ito A, Iwasaki S, Yamaguchi M. Polymer Journal. 2018;**50**:309

[9] Heggs TG. Polypropylene. Weinheim: Wiley-VCH Verlag GmbH & Co. KGaA; 2012

[10] Nwabunma D, Kyu T. Polypropylene Composites. Hoboken, New Jersey: John Wiley & Sons, Inc. 2007

[11] Busico V, Alfano F, Boone HW, Cipullo R, Stevens JC. US 8,129,487B2. 2012

[12] Chum PS, Swogger KW. Progress in Polymer Science. 2008;**33**:797-819

[13] Stevens JC, Vanderlende DD. US Patent 6,960,635. 2005

[14] Tau LM, Chum PS, Karande S, Bosnyak C . US Patent 6,919,407. 2005

[15] Arriola DJ, Carnahan EM, Hustad PD, Kuhlman RL, Wenzel TT. Science. 2006;**312**:714-719

[16] Kaminsky W. Catalysis Today. 2000;**62**:23e34

[17] Arriola DJ, Carnahan EM, Devore DD, Hustad PD, Roger L. US 7,981,992B2. 2011

[18] Mike Chung TC. Macromolecular Reaction Engineering. 2014;**8**:69-80

[19] Ferreira MJ, Martins AM. Coordination Chemistry Reviews. 2006:250

[20] Maddah HA. American Journal of Polymer Science. 2016;**6**(1):1-11

[21] Galli P, Vecellio G. Progress in Polymer Science. 2001;**26**:1287-1336

[22] Kioka M, Kashiwa N. Mitsui Petrochemical Industries. US 4,952,649. 1990

[23] Mao B, Yang A, Zheng Y, Yang J, Li Z. Beijing Inst. Chem. Eur. Patent 0,258,485. 1992

[24] Mao B, Yang X, Li Z, Yang A. China Petrochemical Corp. Chin. Patent CN 1091748. 1994

[25] Cecchin G, Guglielmi F, Pellicani A, Burgin E. Himont Inc. US Patent 5,286,564. 1994

[26] Invernizzi R, Ligorati F. Enchimica Secondaria S.p.A. US Patent 4,506,027. 1985

[27] Iiskola E, Koskinen J. Neste Oy. US Patent 4,829,034. 1989

[28] Karbasi AK, Leinonen T, Sormunen P. Eur. Patent Appl. 0,627,449 A1. 1994

[29] Goodall BL, van der Nat AA, Sjardijn W. Shell Oil. Co. US Patent 4,329,253. 1982

[30] Goodall BL, van der Nat AA, Sjardijn W. Shell Oil. Co. US Patent 4,393,182. 1983

[31] Goodall BL, van der Nat AA, Sjardijn W. Shell Oil. Co. US Patent 4,400,302. 1983

[32] Goodall BL, van der Nat AA, Sjardijn W. Shell Oil. Co. US Patent 4,414,132. 1983

[33] Terano M, Murai A, Inoue M, Miyoshi K. Toho Titanium. US Patent 4,816,433. 1989

[34] Terano M, Soga H, Kimura K. Toho Titanium. US Patent 4,829,037. 1989

[35] Murai A, Terano M, Kimura K, Inoue M. Toho Titanium. US Patent 4,839,321. 1989

[36] Arzoumanidis GG, Karayannis NM, Khelghatian HM, Lee SS, Johnson BV. Amoco Corp. US Patent 4,866,022. 1989

[37] Arzoumanidis GG, Ka9annis NM, Khelghatian HM, Lee SS, Johnson BV. Amoco Corp. US Patent 4,988,656. 1991

[38] Streeky JA, Bersted BH, Blake JW, Feng D, Hoppin CR, Tovrog BS. Amoco Corp. US Patent 6,201,079. 2001

[39] Ramjoie YJE, Sergeev SA, Vlaar M, Zakharov VA, Bukatov GD. Saudi Basic Indust. Corp. US Patent 7,947,788. 2011

[40] Zuideveld MA, Sainani JB, Vimalkumar MP. Saudi Basic Indust. Corp. Eur. Patent Appl. 2,837,634. 2015

[41] Smith GM, Amata RJ, Tirendi CF, Band EI. Akzo Nv. US Patent 5,262,573. 1993

[42] Epstein RA, Wallack WT. Akzo Nv. US Patent 7,504,352. 2009

[43] Albizzati E, Giannini U, Morini G, Smith CA, Ziegler R. Ziegler Catalysts: Recent Scientific Innovations and Technological Improvements

[44] Fink G, Mülhaupt R, Brintzinger HH, editors. Ziegler catalysts: recent scientific innovations and technological improvements. Berlin: Springer-Verlag; 1995. p. 413

[45] Albizzati E, Barbè PC, Noristi L, Scordamaglia R, Barino L, Giannini U, et al. Himont Inc. Eur. Patent 0,361,494. 1989

[46] Morini G, Cristofori A. Montell North America Inc. Eur. Patent 0,728,724. 1996

[47] Barino L, Scordamaglia R. Macromolecular Symposia. 1995;**89**:101

[48] Morini G, Balbontin G, Gulevich YV, Kelder RT, Duijghuisen HPB, Klusener PAA, et al. Montell Tech. Corp. PCT Int. Appl. WO 2000063261. 2000

[49] Cecchin G, Morini G, Pelliconi A. Macromolecular Symposia. 2001;**173**:195

[50] Gao M, Lui H, Li Z, Wang J, Yang J, Li T, et al. China Petroleum Chem. Corp. US Patent Appl. 20050239636. 2005

[51] Guidotti S, Morini G, Esposito S, Mignogna A, Pater JTM, Piemontesi F, et al. US Patent Appl. 2014/0046010. 2014

[52] Guidotti S, Piemontesi F, Pater JTM, Morini G. US Patent 8,829,126. 2014

[53] Severn J, Robert L, Jones JR. In: Ray Hoff., editors. Handbook of Transition Metal Polymerization Catalysts. 2nd ed. John Wiley & Sons, Inc.; 2018. p. 257

[54] Chen L, Leung TW, Tao T. US Patent 8,288,585. 2012

[55] van Egmond JW. PCT. Int. Patent Appl. 2015/081254. 2015

[56] Chen L, Leung TW, Roth GA, Tao T, Gao K. US Patent 8,507,717. 2013

[57] Gullo MF, Roth GA, Leung TW, Williams CC. US Patent Appl. 2013/0053525. 2013

[58] Coalter III JN, Chen L, Williams CC. US Patent Appl. 2013/0338321. 2013

[59] Leinonen T, Denifl P. Borealis Polymers Oy. Eur. Patent 1,273,595. 2003

[60] Rönkkö H-L, Knuutilla H, Denifl P, Leinonen T, Venäläinen T. Journal of Molecular Catalysis A: Chemical. 2007;**278**:127

[61] Abboud M, Denifl P, Reichert K-H. Macromolecular Materials and Engineering. 2005;**290**:1220

[62] Härkönen M, Seppala JV. Macromolecular Chemistry. 1991;**192**:721

[63] Chang M, Liu X, Nelson PJ, Munzing GR, Gegan TA, Kissin YV. Journal of Catalysis. 2006;**239**:347

[64] Härkönen M, Seppälä JV, Väänänen T. In: Keii T, Soga K, editors. Catalytic Olefin Polymerization. Amsterdam: Elsevier; 1990. p. 87

[65] Proto A, Oliva L, Pellecchia C, Sivak AJ, Cullo LA. Macromolecules. 1990;**23**:2904

[66] Ishimaru N, Kioka M, Toyota A. Eur. Patent 0,350,170. 1990

[67] Chadwick JC, Van Kessel GMM, Sudmeijer O. Macromolecular Chemistry and Physics. 1995;**196**:1431

[68] Chadwick JC. Macromolecular Symposia. 2001;**173**:21

[69] Miro ND, Georgellis GB, Swei H. Exxon Chemical. Eur. Patent 0,743,960. 1998

[70] Chen L, Nemzek TL. Union Carbide Chem. and Plastic Tech. Corp. US Pat 7,141,635. 2006

[71] Peil KP, Neithamer DR, Patrick DW, Wilson BE, Tucker CJ. Macromolecular Rapid Communications. 2004;**25**:119

[72] Cai P, Van Egmond JW, Fedec MJ, Goad JD, Brady III RC, Chen L. US Patent Appl. 2013/0005923. 2013

[73] Resconi L, Cavallo L, Fait A, Piemontesi F. Chemical Reviews. 2000;**100**:1253-1346

[74] Corradini P, Guerra G, Cavallo L. Accounts of Chemical Research. 2004;**37**:231-241

[75] Busico V, Cipullo RJ. American Chemical Society. 1984;**106**:6355-6364

[76] Spaleck W, Antberg M, Rohrmann, J, Winter A, Bachmann B, Kiprof P, et al. Angewandte Chemie International Edition. 1992;**31**:1347

[77] Kaminsky W, Külper K, Brintzinger HH, Wild FRWP. Angewandte Chemie. 1985;**97**:507-508

[78] Ewen JA, Jones RL, Razavi A, Ferrara JD. Journal of the American Chemical Society. 1988;**110**:6255-6256

[79] Coates GW, Waymouth RM. Science. 1995;**267**:217-219

[80] Ewen JA, Elder MJ, Jones RL, Haspeslagh L, Atwood JL, Bott SG, et al. Makromolekulare Chemie, Macromolecular Symposia. 1991;**48-49**:253-295

[81] Wild FRWP, Zsolnai L, Huttner G, Brintzinger HH. Journal of Organometallic Chemistry. 1982;**232**:233

[82] Schnutenhaus H, Brintzinger HH. Angewandte Chemie (International Ed. in English). 1979;**18**:777-778

[83] Brintzinger HH, Fischer D, Mülhaupt R, Rieger B, Waymouth RM. Angewandte Chemie (International Ed. in English). 1995;**34**:1143-1170

[84] Ewen JA. Journal of Molecular Catalysis A: Chemical. 1998;**128**:103-109

[85] Coates GW, Waymouth RM. Oscillating stereocontrol: A strategy for the synthesis of thermoplastic elastomeric polypropylene. Science 1995;**267**:217-219

[86] Bruce MD, Coates GW, Hauptman E, Waymouth RM, Ziller J. American Chemical Society. 1997;**119**:11174-11182

[87] Lin S, Waymouth RM. 2-Arylindene metallocenes: Conformationally dynamic catalysts to control the structure and properties of polypropylenes. Accounts of Chemical Research. 2002;**35**:765-773

[88] Hu Y, Carlson ED, Fuller GG, Waymouth RM. Macromolecules. 1999;**32**:3334-3340

[89] Petoff JLM, Agoston T, Lal TK, Waymouth RM. Journal of the American Chemical Society. 1998;**120**:11316-11322

[90] Babkina ON, Bravaya NM, Nedorezova PM, Saratovskikh SL, Tsvetkova VI. Journal of Catalysis. 2002;**43**:341-350

[91] Bravakis AM, Bailey LE, Pigeon M, Collins S. Macromolecules. 1998;**31**:1000-1009

[92] Nejabat G-R, Nekoomanesh M, Arabi H, SalehiMobarakeh H, Zohuri G-H, Omidvar M, et al. Journal of Polymer Science, Part A: Polymer Chemistry. 2013;**51**:724-731

[93] Chien JCW, Iwamoto Y, Rausch MD, Wedler W, Winter HH. Macromolecules. 1997;**30**:3447-3458

[94] Przybyla C, FinkG. Acta Polymerica. 1999;**50**:77–83

[95] Thomann R, Thomann Y, Mülhaupt R, Kressler J, Busse K, Lilge D, et al. Journal of Macromolecular Science, Part B: Physics. 2002;**B41**:1079-1090

[96] Tynys A, Eilertsen JL, Seppala JV, Rytter E. Journal of Polymer Science, Part A: Polymer Chemistry. 2007;**45**:1364-1376

[97] Weng W, Dekmezian A, Markel EJ, Gadkari A, Dekoninck J-M. US Patent 6,750,307B2. 2004

[98] Nakamura A, Ito S, Nozaki K. Chemical Reviews. 2009;**109**(11): 5215-5244

[99] Huang H-h, Zhang C-h, Qin Y-w, Niu H, Dong J-y. Chinese Journal of Polymer Science. 2013;**31**(4):550-562

[100] Amin S B, Marks T J. Angewandte Chemie, International Edition. 2008;**47**:2006-2025

[101] Makio H, Kashiwa N, Fujita T. Advanced Synthesis and Catalysis. 2002;**344**:477

[102] Tian J, Coates GW. Angewandte Chemie, International Edition in English. 2000;**39**:3626

[103] Diamond GM, LaPointe AM, Leclerc MK, Longnire J, Nava-Salgado V, Shoemaker JAW, et al. Symxy Tech. US Patent 7,256,296. 2007

[104] Boussie TR, Brümmer O, Diamond GM, Goh C, LaPointe AM, Leclerc M, et al. Symxy Tech. US Patent 7,241,715. 2007

[105] Buscio V, Cipullo R, Ronca S, Budzelaar PHM. Macromolecular Rapid Communications. 2001;**22**:1405

[106] Boussie TR, Diamond GM, Goh C, Hall KA, LaPointe AM, Leclerc MK, et al. Angewandte Chemie,

International Edition in English. 2006;**45**:3278

[107] Soares JBP, Simon LC. Chapter 8: Coordination Polymerization. In: Handbook of Polymer Reaction Engineering. Weinheim: Wiley-VCH Verlag GmbH & Co. KGaA; 2005. ISBN: 3-527-31014-2

[108] Arriola DJ, Carnahan EM, Devore DD, Hustad PD, Huhlman RL, Wenzel TT. WO 2005/090425 A1. 2005

[109] Arriola DJ, Carnahan EM, Devore DD, Hustad PD, Huhlman RL, Wenzel TT. WO 2005/090427 A2. 2005

[110] Hustad PD, Kuhlman RL, Shan CLP, Poon BC, Roof GR, Stevens JC, et al.. WO 2005/090426 A1. 2005

[111] Alfano F, Boone HW, Busico V, Cipullo R, Stevens JC. Macromolecules. 2007;**40**:7736-7738

[112] Wang J, Liu X, Jia J, Chen X, Zhu B. Advances in Materials Research. 2011;**396-398**:8-12

[113] Ijpeij EG, Coussens B, Zuideveld MA, van Doremaele GHJ, Mountford P, Lutz M, et al. Chemical Communications. 2010;**46**:3339-3341

[114] Ijpeij EG, Windmuller PJH, Arts HJ, van der Burgt F, van Doremaele GHJ, Zuideveld MA. DSM, WO 2005/090418 A1. 2005

[115] Windmuller PJH, van Doremaele GHJ. DSM, WO 2005/005496 A2. 2005

[116] Ijpeij EG, Zuideveld MA, Arts HJ, van der Burgt F, van Doremaele GHJ. DSM, WO 2007/031295 A1. 2007

[117] Dorini M, Mei G. Chapter 17: Spherizone technology. In: Cavani F, Centi G, Perathoner S, Trifiró F, editors. Copyright 2009.Sustainable Industrial Processes. Weinheim: Wiley-VCH Verlag GmbH & Co. KGaA. ISBN: 978-3-527-31552-9

[118] Covezzi M, Mei G. Chemical Engineering Science. 2001;**56**:4059-4067

Chapter 3

Thermal Conductivity of Polypropylene-Based Materials

Antonella Patti and Domenico Acierno

Abstract

In this work, the authors aimed to provide an overview about the thermal conductivity of polypropylene, of its related compounds and the main methods of measurement. The growing spread of polypropylene in the industrial world together with the increasing demand of thermally conductive plastics represented the driving force of studying the heat transport in the polypropylene, and of recent progress and development of the thermal conduction in polypropylene-based materials. At regard, the common approach has been devoted to fill the polymer with thermally conductive materials: metallic, carbon based, ceramic and mineral fillers have been taken into account depending on the need to preserve electrical insulation, lightweight, production increasing or cost saving in the final compositions. Different parameters have been considered in order to optimize the ultimate thermal performances in the realized composites: (i) filler dispersion, (ii) filler/matrix and filler/filler interactions. The introduction of functional groups on the filler surface or in the polymer chain has been tested for acting on the dispersion and on the interfacial interaction. Then, hybrid materials, consisting in two particles different in size and shape combined with the attempt to realize a synergistic effect and to support a conductive network in the matrix, have been investigated.

Keywords: polypropylene, heat transport, measurement methods, enhancement, composites

1. Introduction

Polypropylene (PP), once discovered, immediately has been appreciated for its remarkable and various qualities like low density (the lowest compared to all thermoplastics), excellent chemical and corrosive resistance, dimensional stability, recyclability, flexibility, good processability, and low cost. All these advantages have contributed to make it ideal for a wide range of applications and to be processed thought many converting methods [1].

Injection molding is a very common used technique to work polypropylene, so much to cover about a third of its consumption [2]. In automotive, injection molding represents the main process for realizing vehicle components and PP is one among the conventional used thermoplastic materials. In the car, polypropylene is applied in the interior trim components as pillars, glove box, console bins and console housings, or exterior parts as air inlet panels and wheelhouse liners, or again under the hood for producing a fuse block cover, radiator, and coolant reservoir [3]. Extrusion, instead, is the single most popular process for forming PP in fibers, filaments, sheets and other products, depending on the die profile [4]. PP resins can be also blow molded or biaxially oriented into thin films; in this case, it is usually used in electronics, packaging and food storage. Many products, such as printings films, magnetic films, decorations films, heat-sealing films, metal electronic deposition films are realized by biaxially oriented PP (BOPP) [5].

Moreover, polypropylene along with others thermoplastics as polyvinylidene fluoride (PVDF), polytetrafluoroethylene (PTFE), polyethylene (PE), polycarbonate (PC), polyphenylene sulphide (PPS), polyphenylene oxide (PPO), has also been considered in heat exchanger applications in order to replace heavier and expensive metals strongly subjected to corrosion problems. Different models of heat exchangers have been also studied as shell and tube, plate, finned tube, and hollow fibers [6]. Yet, the overall heat transmission coefficient in polymer heat exchangers had been lower with respect to that obtained by metals and different approaches have been suggested to improve the heat transport in this field. The first was concerned with the reduction in thickness wall by requiring a new type of design; the second consisted in an increase of the thermal conductivity (TC) of the polymer resins [7].

Unfortunately, most polymers are thermally insulators and an augment of their TC represents one of the challenges of the recent scientific research for the coming benefits in the emerging heat exchangers applications, and also in the electronic and automotive technologies. In fact, plastics are one of the common materials for covering and protecting the electronic components or a better alternative to produce lighter vehicles with reduced fuel consumption. For example, in electronic assembly, miniaturization and higher power density have been experimented in order to achieve more advanced performances: a single chip has been made up of lots of transistors. This led to an intensification of power and a significant heat flow that can negatively affect the lifetime and reliability of the device, if it is not opportune dissipated away in time [8]. In the modern car engine, the electronic control unit (ECU) involves both hardware and software, required to perform the functions as mixture formation, combustion and exhaust gas treatment. In order to accommodate the more stringent standard imposed by the European Community on the emissions of pollutants from cars (Euro 6 standards), new kinds of control unit platform with high capability and reduced weight and size have been studied. Due to an inevitable increase of power density, thermal management has become an important aspect linked to the ability to easily dissipate the heat [9]. In an internal combustion engine, an elevated heat transfer affects performance, efficiency and emissions because it reduces the average combustion gas temperature and pressure, and the work per cycle transferred to the piston, for a given mass of fuel within the cylinder [10]. In automobile components, an improved TC of the constituent materials for engine cover, radiator, and coolant reservoir could be convenient to promote the faster diffusion of the heat, in order to avoid the overheating of the overall system and compromise the respective performances. In injection molding or extrusion, the greater would be the TC of the processed materials, lower would be the heating or cooling time and the operating costs of the overall processes. Moreover, during the cooling phase of these processes, sudden variation of TC could determine shrinkage, stress, delamination or voids in the final products [11].

In improving the heat conduction of the plastics different efforts have been spent in the scientific research with the addition of fillers characterized by a TC superior to the matrix [12–14].

In this work, the authors have focused the main attention on literature studies involving TC of polypropylene and the factors on which it depends as the crystallinity, testing temperature and pressure, polymer chain orientation, and molecular weight. Following, the effect of introducing fillers on the heat transport ability of the PP matrix has also been illustrated. Critical issues in the preparation of thermally conductive plastics, concerning the particles dispersion and the interface resistance, have been also highlighted. Finally, filler functionalization and hybrid materials have been presented as methods to support the heat conduction in the PP-based composite systems.

2. Thermal conductivity of polypropylene

2.1 Definition of thermal conductivity in polymers

In general, heat transfer takes place through three different mechanisms: convection, conduction, irradiation.

In a solid, the main mechanism of heat transport is the conduction corresponding to the transfer of particle vibration energy to the adjacent particles without any motion of the matter but exclusively due to collision [12].

In steady state condition, Fourier' law (Eq. (1)) [15] describes the heat conduction across a slab of solid material of surface area A and thickness Y whose sides are set at two different temperatures T_1 e T_0, respectively, in the direction normal to a slab surface (see **Figure 1**):

$$Q = k\,A\,\frac{\Delta T}{Y} \tag{1}$$

In Eq. (1), Q is defined as the heat flow required to maintain the temperature difference $\Delta T = (T_1 - T_0)$ between the two opposite surfaces; Y represents the conduction length path; A is the cross-sectional transfer area, and finally k is the TC of the slab material (in W/mK, SI units).

In the 1932, Debye introduced the concept of mean free path of thermal waves, simply called phonons, for explaining the thermal conduction in a dielectric crystal, in which the electrons are not free to move. In this case, the TC has been described through the following expression, also called Debye equation (Eq. (2)):

$$k = \frac{c_p\,\upsilon\,l}{3} \tag{2}$$

where c_p is the specific heat capacity per unit of volume, l is the phonon mean free path and υ is the average phonon velocity.

Usually, the thermal conduction occurs through these phenomena: (i) by charge carriers as electrons and holes,also defined as energized electron motion; (ii) by phonons, i.e., energy quanta of atomic lattice vibrations due to atom interaction and collision and (iii) by photon conductivity verifiable only in the case of high temperature. Depending on the material's nature, the different phenomena do not always

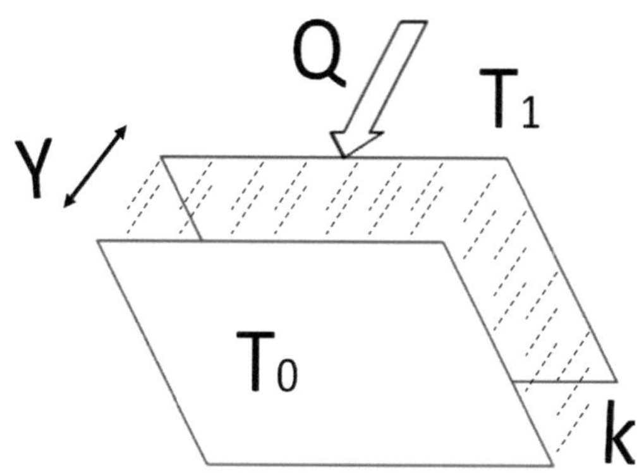

Figure 1.
Heat conduction across a slab of a solid material.

happen simultaneously, but one can dominate over the others. For example, in metals, the electronic contribution exceeds the phonon one; whereas in insulators, phonons contribution prevails over the electrons one [11–14]. Polymers are thermal insulators, and due to defects, grain boundaries, and/or scattering with other phonons the mean free path of phonons (l) is very low and consequently also their TC [16]. For most of thermoplastics, the TC at 25°C falls in the range of 0.11 W/mK (for polypropylene) and 0.44 W/mK (for high density polyethylene) [17].

2.2 Methods for measuring the thermal conductivity of polypropylene-based materials

The classical steady state (SS) and non steady-state (NSS) methods are the two main techniques for evaluating the TC of a material. In the first case, the measurement is carried out after reaching the equilibrium state, while in the second case, the test is performed during the heating phase [18].

The SS mode allows measurements on glass, polymers, insulators, ceramics, metals, composites with an uncertainty of 2–3%; the NSS mode permits to extend the testing also of liquids, gases or powders with an uncertainty up to 10% [12]. However, the first approach is more time consuming and not suitable for shaped samples as concentric cylinder or sphere.

In the SS method, the principle of operation is based on Fourier' law, while in the NSS method, the TC is indirectly evaluated by measuring the thermal diffusivity (α), according to Eq. (3):

$$k = \alpha\, c_p\, \rho \tag{3}$$

Here, c_p and ρ are the heat capacity and the density of the tested material, respectively.

The common apparatuses for evaluating the performance in heat transport of PP and its related compounds are: the "Guarded Hot Plate Method (GHPM)" and the "Heat Flow Meter Method (HFMM)", based on SS approach, or the "Flash Method (FM)" and the "Transient Hot Wire Method (THWM)", instead based on NSS approach.

The differences among these techniques are found essentially in required time for testing and operation mode [19].

The GHPM (**Figure 2a**) is constituted by a hot plate, placed between two samples of the examining material. The outer side of each specimen is in contact with a cold surface. A known heat flow (i.e., the heating power) is applied to one side of the sample and passes through it by establishing a temperature gradient between the two opposite faces of the sample. At steady-state, the TC is evaluated by measuring the difference in temperature and applying the Eq. (1).

The layout of the HFMM (**Figure 2b**) is very close to the GHPM apparatus designed for a single sample, but the first arrangement is faster and more accurate with respect to the second one [18]. The principal difference concerns the replacement of the main heater with heat flux sensor (HFT) [12].

In the FM (**Figure 2c**), a short but intense energy pulse is sent to one side of the sample and the temperature increase is measured on the opposed side in function of time. The thermal diffusivity can be calculated by Parker Formula (Eq. (4)) [20]:

$$\alpha = \frac{1.38\, d^2}{t_{1/2}} \tag{4}$$

where d is the sample's thickness and $t_{1/2}$ is the time required for the signal to reach the 50% of its maximum value.

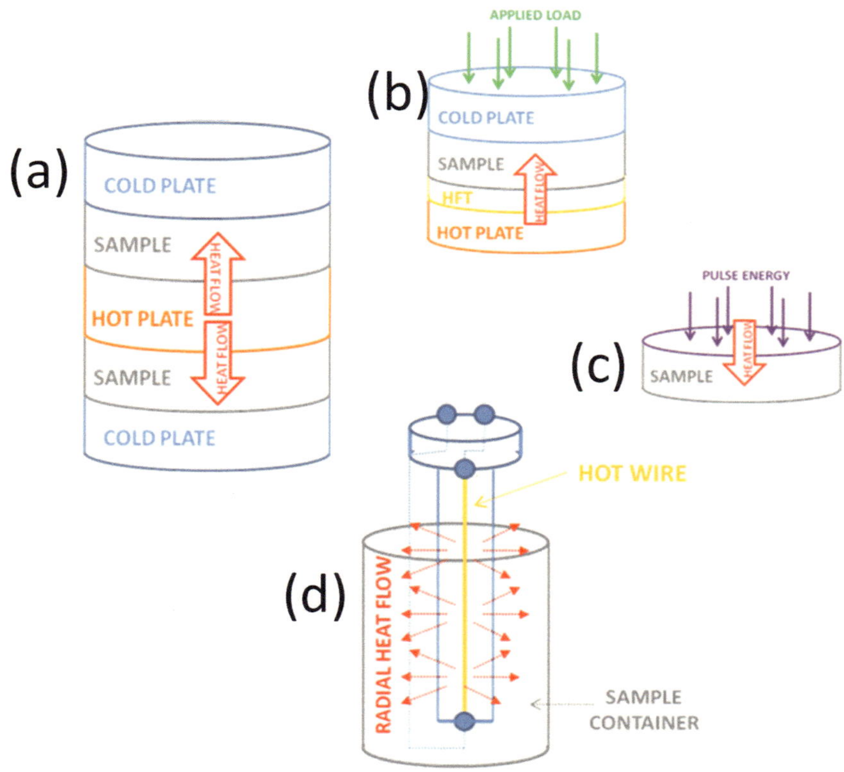

Figure 2.
Schematic representation of TC measuring devices: (a) GHPM; (b) HFMM; (c) FM; (d) THWM.

The THWM (**Figure 2d**) implicates vertical or cylindrical geometry in which a wire, generally of platinum, is crossed by a constant electric current. A radial heat flow takes place around the wire that spreads in the tested sample. The TC is estimated, knowing the temperature profile on time (T(t)) and the heat output by Eq. (5) [21]:

$$k = \frac{q}{4\pi(T(t) - T_0)} ln\left(\frac{4t\alpha}{r^2 C}\right) \tag{5}$$

in which ln C is the Euler's constant, q is the applied thermal flow, r is the wire radius and T_0 is the cell temperature.

Finally, also differential scanning calorimetry (DSC) has been considered as a technique for measuring the TC of solid materials. The analysis has been performed by incorporating a temperature sensor [22] or a made in-house accessory into the common apparatus [23]. It is possible also utilized the standard DSC machine, without any special modification or calibration, by setting a specific temperature-time profile and recording the dynamic response of the sample [24].

2.3 Factors affecting the thermal conductivity of polypropylene

The TC of polymers is affected by several factors as crystallinity, chemical constituent, bond strength, molecular weight, side pendent groups, defects or structural faults, processing conditions and temperature [14].

Anderson [16] reported that the TC of polymers decreased as the disorder increased: imperfections, by decreasing the order of the molecular structure, caused a large phonon scattering that reduced the heat transport. Since the polymeric structure order of the amorphous is lower than of the crystalline, the related thermal behaviour of the former has been expected to be lower compared to the latter, and also TC temperature dependence has changed in different ways depending on substance state. In details, below the glass temperature, as the temperature grew up, the TC of the amorphous remained the same or increased with temperature (probably for the effect of raising chain mobility), while for crystalline the TC initially remained the same and then diminished. Probably, in this second case, a decrease or/ and breakup of the crystalline portions have been promoted by higher temperatures after which the conductivity of the amorphous has been risen. These considerations were confirmed by studies of Bashirov et al. [25] and Osswald et al. [26],developed not only on PP, but also on high-density polyethylene (HDPE), low-density polyethylene (LDPE) and other polymers. On the contrary, an opposite trend of the TC of PP against temperature was found by dos Santos et al. [11]. In their work, the authors measured the TC of semi-crystalline and amorphous polymers starting from room temperature and going up to melting temperature (for semi-crystalline polymers) or glass transition (for amorphous polymers). Results showed that initially, as the temperature rose between 25 and 125°C, the TC of PP slight decreased from 0.25 to 0.15 W/mK; then, it underwent a sudden increase reaching a peak of approximately 0.47 W/mK during the melting process, and finally it decreased. In the temperature range of 2–100 K the TC of polypropylene was evaluated by Choy et al. [27]. In the case of isotropic crystalline conditions, an increasing trend that exhibited a maximum near 100 K was detected; then, when the sample was extruded and a marked anisotropy of TC was induced, the heat transport resulted in an order of magnitude increase in the extrusion direction. Finally, a nearly linear temperature dependence without any detectable plateau of the TC of PP copolymer was observed by Barucci et al. in the range of temperature from 0.1 to 4 K [28].

Most of the TC measurements of polymers have been carried out at atmospheric pressure, which is far from the operating process conditions. At regard, some measurements have been reported in order to verify the effect of pressure on TC of polymers. Dawson et al. [29] measured the thermal parameter for polypropylene at pressures of 20, 80 and 120 MPa over the temperature range from 250 to 50°C. For each pressure, the isobaric curve of TC as a function of temperature showed a "Z shaped", probably attributed to a phase transition (crystallization during cooling). Beyond an increase encountered during this phase, the TC remained fairly constant with the temperature. At an equal temperature, an increase of TC was always verified with pressure, approximately of 20% going from 20 to 120 MPa. Andersson et al. [30] performed measurements of heat conduction of PP at 300°K by changing pressure in the range between 0 and 37 bar. They concluded that, when the pressure was exercised on the sample, a stress in the longitudinal direction was generated, greater than in the radial direction, by leading to anisotropy of the properties in the tested material. Experimental data, related both to atactic and isotactic PP, demonstrated that the TC increased strongly with pressure with a continuous change in the slope of curve until it reached an asymptotic value.

From the above, the actualizing the anisotropy in a sample has implied an influence of its thermal transport behaviour. In fact, when the orientation has been induced in the polymer, its TC became higher in the direction of a molecular orientation and lower in a direction normal to the orientation [31]. This attitude was confirmed in the case of injection molding and extrusion [32] or foaming [33] processes during which a macromolecular orientation of polymer chain was inferred.

Finally, a characterization of the heat transport directly on the melted PP has been carried out because of in a common process the material was usually in the molten state. In fact, in a solid state the TC of semi-crystalline thermoplastics was greater with respect to melt state due to an increase of density upon the solidification. At regard,

Polypropylene	TC (W/mK)
Han et al. [14]	0.11
Goswami et al. [35]	0.12–0.17
Maier et al. [36]	0.17–0.22
Tripathi [4], Guo et al. [37], Birley [38]	0.22

Table 1.
Data reported in literature on TC values of polypropylene.

investigations about the effect of hydrostatic pressure, temperature, and chemical structure on the thermal conduction of melted PP have been performed [26]. These studies confirmed that, as the hydrostatic pressure increased on melt state, also the TC of thermoplastics in general, and of PP in particular, increased for a reduction in free volume. Furthermore, the TC of PP in molten form was not significantly affected by temperature but it appeared to be a complex function of the molecular weight distribution and possible long chain branching [34]. Generally, an increment of the TC of polymers by raising the molecular weight was verified since a larger number of energy transactions took place in a substance with shorter chains [16].

In **Table 1** the TC values for polypropylene are shown.

3. Enhancement of thermal conduction in polypropylene-based materials

3.1 Thermally conductive fillers in polypropylene

In general, the common approach for enhancing the thermal transport behaviour of plastics foresees the addition of thermally conductive particles. By balancing in the polymeric resin the filler content and type, it is possible to obtain the desired features in the final products. Yet, the use of an extremely high percentage of reinforcement (approximately more than 30% in vol.), is needed to achieve the TC values in the composites, required for the modern technologies. This quantity represents a real challenge for the processability of the material and makes difficult or impossible extrusion and injection molding processes [14]. In the last few decades, great attention has been devoted to polymeric nanomaterials, born from the introduction into the matrix of filler having at least one dimension in the order of 1–100 nm. Based on the geometric characteristics, three groups of nanosize particles are distinguished: one-dimensional (nanotubes and nanofibres), two-dimensional (layered minerals), three-dimensional (spherical particles). Small size and large surface area (for a given volume) of nanofiller are considered the key factors for the development of exceptional and unexpected properties with respect to macroworld as in the fields of mechanical properties, barrier resistance, flame retardancy, scratch/wear resistance, as well as optical, magnetic, TC and electrical properties [17]. Thermally conductive fillers can be divided into three categories: metallic powders, ceramic particles and carbon-based materials, and have been chosen depending on the needs to act both on the heat and current transport. For example, by adding carbon-based or metallic particles, the final compounds earned not only in terms of the thermal conduction but also in terms of the electric one; yet, metallic particles, having high specific gravities, could not be applied in the case of the lightweight target and carbon-based reinforcements have been preferred. Conversely, the introduction of ceramic fillers allowed acting on heat transfer of the neat matrix without compromising the electrical insulation of the starting material [39].

The thermal and electrical behaviour of PP matrix filled with two of copper particles (in micron dimensions), was investigated by Boudenne et al. [40]. Results, compared in terms of fillers size and volume fraction, highlighted the stronger heat transport ability of the formulations including the smaller particles.

A similar study was conducted by Cheewawuttipong et al. [41] by adding into PP two types of boron nitride (BN) with different particle size in micron. The mechanical features (storage and loss modulus) and thermal ones (thermal conductivity, melting and crystallization temperature) were analyzed. They found that the TC increased, according to the BN content and the larger size of the filler.

The application of conductive polymer composites (CPC) in manufacturing a tubular heat exchanger for fluid heating was presented by Glouannec et al. [42]. The materials of the extruded tubes were obtained by blending the insulating thermoplastic polymer (PP) with conductive filler like carbon black (CB) or carbon fibers (CF). The experimental testing showed that the heat conduction of the CPC heat exchanger was improved by a factor of 2 for the filler volume fraction of 25%. For similar applications, Qin et al. [43] realized graphite (GP) enhanced PP hollow fiber heat exchanger. Results showed that the addition of GP fairly improved the crystalline, thermal stability and conductivity of the PP. The overall heat transfer coefficient for a filler content of 15.0 wt% became five times bigger than that of pure PP-based one.

Among the carbonaceous fillers, given the extremely high intrinsic TC, the carbon nanotubes (CNTs) have been also considered [44]. The TC of the PP-based nanocomposites grew up with the CNTs content in particular above 160°C, but the rising trend not seemed to be dramatic as the increase in electric conductivity.

However the advantages of using minerals in many polyolefin applications should not be overlooked. It has been successfully demonstrated that talc, calcium carbonate, and magnetite led to an improvement of the productivity and of physical properties, and to a reduction of costs since a portion of the polymer has been replaced by lower-cost material [45, 46]. Moreover, due to a higher TC of minerals with respect to the matrix, their addition also brought an increase of heat transport in the ultimate compounds. The introduction of calcium carbonate ($CaCO_3$) nanoparticles in isotactic PP was studied by Vakili et al. [47], verifying the particle effect on crystallization and heat transport behaviour of the ultimate compounds. The TC was increased from pure PP by 64% for the sample containing the highest filler loading (15 phr). The authors stated that their results were better than that obtained with carbon nanofiber (CNF) in the PP matrix [48], for which at 0.08 volume fraction this increase has been almost 45%. Weidenfeller et al. [49] added different micro-size particle of magnetite in various proportions to a standard polypropylene copolymer. The TC increased from 0.22 to o.93 W/mK for a filler content of 44% in vol. of magnetite, nearly independent of the particle size and distribution; whereas the electrical resistivity decreased more than seven orders of magnitude. Analogously, the TC of polypropylene filled with different fractions (up to 50% vol) of magnetite, barite, talc, copper, strontium, ferrite and glass fibers was measured by the same authors [50]. Despite the highest intrinsic TC of copper, surprisingly, the best result was obtained with talc. Using the model conductivities, according to Schilling and Partzsch, the interconnectivity of the conducting phase has been evaluated. Calculations showed that the interconnectivity for talc was particularly greater than that of copper probably due to the alignment of the talc lamellas in the polymeric resin into the flow direction during injection molding; for copper no contact among particles has been found. The authors concluded that the thermal transport could not be explained solely by the difference in the property of the materials but taking into account also the microstructure.

Fillers	Category	Typical shape	TC (W/mK)
Allumium nitride (AlN)	Ceramics	Roughly spherical	320 [30]
Alluminium oxide (Al_2O_3)	Ceramics	Spherical or platelets	42 [8]
Barium titanate ($BaTiO_3$)	Ceramics	Grains	6.2 [8]
Boron nitride (BN)	Ceramics	Hexagonal platelets	3 (through plane) 300 (in plane) [41]
Zinc oxide (ZnO)	Ceramics	Pseudo-spherical	60 [8]
Carbon black (CB)	Carbon-based	Pseudo-spherical	6-174 [39]
Carbon nanotubes (CNTs)	Carbon-based	Cylindrical shells	2000–6000 (on axis) [39]
Carbon fiber (CF)	Carbon-based	Elongated	600 [48]
Graphene (GN)	Carbon-based	Sheets	5000–6000 [39]
Graphite (GP)	Carbon-based	Platelets	100–400 (on plane) [40]
Copper (Cu)	Metals	Pseudo-spherical	389 [40]
Calcium carbonate ($CaCO_3$)	Mineral	Pseudo-spherical	4.5 [47]
Magnetite	Minerals	Grains	9.7 [49]
Talc	Minerals	Lamellar	1.76 (through plane) 10.69 (in plane) [50]

Table 2.
TC values of common fillers added into polypropylene.

A list of the used fillers to increase the thermal transport behaviour of PP and their intrinsic TC are given in **Table 2**. These particles are added in the hosting matrix by prevalently melt blending: one among the classical processes for compounding, particularly preferred in an industrial context.

3.2 Critical aspects for optimizing the thermal conduction in composites

According to the rule of mixture (Eq. (6)), each phase contributes to final compounds properties proportionally to its volume fraction:

$$p_c = \sum_{i=0}^{n} p_i \, \phi_i \tag{6}$$

where p_c is composite property, p_i is the intrinsic property of each specific constituent and ϕ_i is their volume fraction in a composite system.

However, the experimental evidences confirmed that the values of the TC in the designed composites were lower than those evaluated by applying the simple rule of mixing model and the intrinsic TC of each constituent. This meant that nanoscale unique properties could not be reproduced easily on the macroscale probably for non-continuum effects at filler-polymer interfaces [51].

In fact, besides the filler type and its content, particles shape and size, their orientation and dispersion in polymeric matrix, the adhesion between the filler and the matrix and the thermal properties at the interface symbolized the acting parameters to optimize the desired feature [12].

The interfacial resistance is also known as "*Kapitza Resistance*" from the name of the discoverer that in 1941 found the temperature discontinuity at the interface between metal (copper) and liquid (superfluid helium). It represents a barrier to heat flow at the interface between two phases, due to possible weak contact and differences in the phonon spectra [52]. The thermal contact resistance is attributed to phonons losses during heat transmission from one medium to another and is qualitatively described by two limiting models: acoustic mismatch model (AMM) and the diffuse mismatch model (DMM), representing the upper bound and lower bound of thermal boundary resistance, respectively. The former assumes that all phonons propagate as a planar wave, so the transmitted energy is related to different acoustic impedances of two materials. Instead, the latter assumes that all phonons, colliding into the interface, are randomly scattered. If the acoustic impedances between materials are very different, the phonons scattering will contribute to further increase the thermal resistance [53].

In addition to Kapitza resistance, in a composite a further resistance to the heat transport, involving the solid-solid interface, i.e., particle-particle interface has to be considered. Two solid bodies, which are apparently in contact, actually touch together only in a few points due to their roughness and geometrical defects. In this case, when heat flows normally from a hotter body to colder one, interstitial spaces between contact points limit the thermal conduction that takes place only between effective connections. The physical consequence is a constriction of flux lines, responsible of the heat resistance best known as "thermal contact resistance" [54].

In many previous works, it has been reported that two essential components for an optimal effect of the reinforcement in a composite are filler dispersion and orientation. A homogeneous filler dispersion in the resin led to a consistent load transfer from the matrix to particle and in the realization of a conductive network for electrical and thermal energy. Furthermore, if the particles were oriented in the direction of applied force or heat, their full potential could be achieved both in terms of load and energy transport. However, an efficient dispersion or/and a perfect alignment of the inorganic filler, particularly for nano-sized, in the matrix represented a real challenge during the process due to van deer Waals interactions between individual particles that push to the aggregation and randomly disposition [55]. Yet, in the case of heat transport, the enhancing the filler dispersion not always implied exceptional thermal performance of respective composites, because an increase in the matrix filler interfaces, and consequently in the phonon scattering, could be promoted. In particular for nanocomposites, the large surface area of nano-size particles, by maximizing the touching between particle and polymer, could increase exactly the interfacial resistance. Thus, particles perfectly dispersed in the matrix with a minimized interfacial resistance would not seem anyway the best solution in the perspective of an advantageous heat transport because in this situation the particles would exchange heat only with the matrix and not among themselves. Since the mean phonon free path of the polymer is less than that of the particle, the efficiency of the thermal conduction would not be satisfactorily achieved except by a network of interconnected particles that carry heat among themselves [14].

Additionally, in the composites the conductive pathway can be build also by particle agglomerates if a sufficiently reduced "*mean interparticle distance*" (MID) has realized among them. So, in the case of poor dispersion and low filler loading, the isolated agglomerates resulted not effective as well the dispersed ones. In other words, the thermal network could also be get among the same aggregates that behave like larger-size conductive particles without interfaces by actualizing a sufficiently reduced MID. In this perspective, the aspect ratio of particles became an interesting and important parameter to be considered because bigger was the length of particles, lower could be the filler/matrix interface and scattering of phonons, and greater was the TC value in the final system. Nevertheless, at an equal filler percentage, if on one side the micro-dimension of the particles compared to nano-size could be favorable for reducing the interfaces, on the other side it contributed to an increase in MID [13].

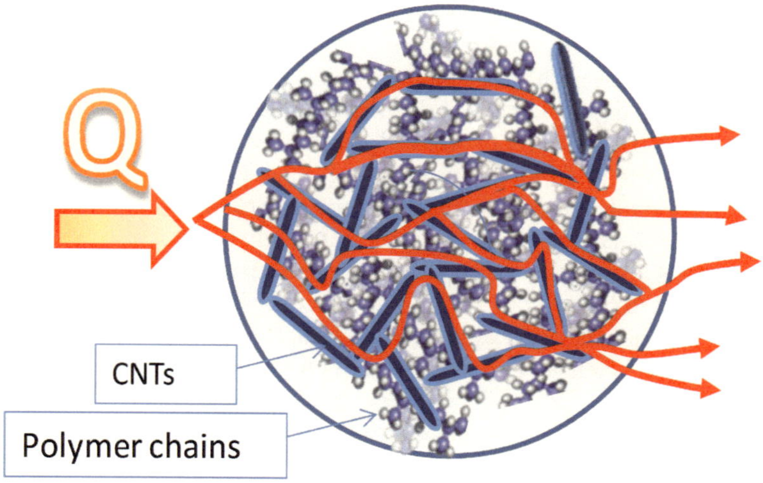

Figure 3.
Schematic of the heat conduction along particles forming a percolated network.

In conclusion, in order to successfully employ the thermal conductive particles in composites for heat management applications, the realization of a conductive pathway through the particles (shown in **Figure 3**) has to be attained through optimal filler dispersion, good interfacial adhesion between filler-matrix and properly contact between particles.

3.3 Effect of functionalization

The functionalization, i.e., the introduction of functional groups on filler surface, obtained by covalent or non-covalent means, is a common approach widely adopted for changing the chemistry of inert inorganic particles. This approach is considered a successful tool for improving the compatibility and wettability between the two phases of the composite system, and affecting positively the dispersion and the interfacial resistance. Although few works are carried out on this way, it is not excluded that between functional groups may arise interactions, stronger than van deer Waals forces, which, binding more particles between them, also promote a reduction in the contact resistance.

Patti et al. [56] studied the effect of filler functionalization on dispersion and TC of PP-based composites by adding, in the same resin grade, three different chemically treated surfaces (amino-and carboxyl-functionalized, and pristine one) CNTs. A cubic polynomial trend of the TC as a function of the filler concentration was found, for all three families of nanocomposites. This behaviour was attributed to the occurrence of competitive dispersion/agglomeration phenomena which, affecting the mutual distance among filler aggregates/agglomerates (MID), influenced the formation of thermal paths.

CNTs, aluminum flakes (Al-flakes), and a commercially available Al-CNT powder (embedded CNTs within Al-flakes), were used to prepared PP-based composites. The characterization of these compounds in terms of mechanical and thermal tests indicated that the crystallization and decomposition temperature as well as the TC and tensile modulus of PP/Al-CNT were over than to the PP/CNTs and PP/Al-flakes composites. It was hypothesized that free CNTs, produced during the preparation of PP/Al-CNT systems, played an important role in forming a conductive bridge among Al-CNT particles and in generating synergistic effects [57].

Muratov et al. introduced two different types of surface treated BN into the PP matrix: pristine and covered with silane coupling agent (–3-amino-propyl-3-ethoxy-silane-APTES). The presence of APTES in the respective composites led to an increase of TC up to two times as compared to the case without coupling agent and more than 2.5–4 times as compared to pure PP [58].

Thus, if the first method for improving the compatibility filler-matrix is the modification of the filler surface by using functional groups or coupling agents, the other method consists in the modification of the matrix by grafting reactions of different chemical groups. For example, in the case of polyolefins, maleic anhydride is commonly grafted to polypropylene for improving the filler-matrix interface through both physical and chemical interactions, as covalent linkages and hydrogen bonds [59]. Che et al. [60] functionalized the surfaces of micrometric BN with a thin layer of polydopamine coatings (f-BN). Then, maleic anhydride grafted PP (PP-g-MA) was also employed as the compatibilizer in the compounds for helping the covalent bonds with polydopamine. The highest value of the TC was recorded for the prepared mixtures containing the compatibilizer and functionalized filler. On the contrary, the formulations containing pristine BN possessed a higher TC with respect to that prepared with modified one. Probably, the coating of polydopamine layers increased the polarity and hydrophilicity of fillers, which would be less compatible with the non-polar and hydrophobic PP matrix, resulting in voids at the interface and filler aggregation, and inducing a strong interruption of the thermal paths. Szentes et al. [61] utilized novel types of coupling agents (olefin-maleic anhydride copolymer-based) as a compatibilizer in CNTs/PP systems. They found that neither the chemical structure of coupling agent nor the application methods (masterbatch and impregnation) have been conclusive on the heat conduction of prepared compositions.

Although the experimental evidences seem to be lesser and to be explored in greater detail, it should be emphasized that the increasing the compatibility between filler and matrix not always a positive effect on heat transport has been verified. Compatibilized samples sometimes display slightly lower TC with respect to non-compatibilized systems, probably due to a polymeric wrapping around particles that hinders the direct contact among themselves and, as a consequence, the thermal transmission [62]. Patti et al. [63] found that the addition of PP-g-MA significantly reduced the contribution of included CNTs to the ultimate thermal transport properties of the corresponded mixtures, from +42.2 to +19.6% in presence of neat CNTs and from +47.7 to +11.7% for systems containing amino functionalized CNTs.

3.4 Hybrid materials

Nowadays, recent scientific attention has been devoted to hybrid material defined as "a combination of two or more materials in a predetermined geometry and scale, optimally serving a specific engineering purpose". A hybrid has been conceived as a mixture of two or more raw materials with own intrinsic properties and different shape and size ("A + B + shape + scale") to enhance or diminish physical, mechanical, thermal and electrical properties, as stiffness or strength, and also to manipulate the percolation limit [64]. In fact, the difference in dimension and geometry allows to maximize the packing density and to favor a greater connection among the various particles. The final properties of hybrids could be a combination between the features of the individual constituents or a result of synergism, intended as a total effect greater than the sum of each component, due to a joint action between fillers, supplied by distinct geometry of shape, aspect ratio as well distribution [65].

King et al. [66] analyzed the effect of single carbonaceous fillers (CB, GP, CNTs) and their combination on the TC of PP in the light of a potential market of fuel cell bipolar plates. The association of these fillers in the matrix appeared always more effective in improving the tested feature compared to the potential of each single species. In the case of hybrid formulations, the best result was obtained by mixing together the three particles, probably for the formation of linkages among them. In the case of composites including an individual filler, the highest TC value (28 W/mK) was achieved with 80% in wt. of GP and was anyway higher than the desired TC required for bipolar plates (20 W/mK). Krause and Potschke [67] investigated another useful combination of carbon-based fillers (CNTs, CF, graphite nanoplatlets-GNP) in PP by analyzing both the thermal and the electrical conduction of melt blended compounds. Ren et al. [68] obtained the simultaneous enhancement on thermal and mechanical properties of PP composites by adding graphite platelets (GPs) and graphene sheets (GSs). Yao et al. [69] reported the enhancement on the TC and dielectric properties of PP composites due to the synergistic effect of the introduced nano-sized aluminum oxide (Al_2O_3) and micro-sized barium titanate ($BaTiO_3$). At a maximum of filler content (50% in vol.) the $BaTiO_3$/PP and Al_2O_3/PP composites exhibited the same increment in TC approximately equal to 100%, while for $BaTiO_3$/Al_2O_3/PP systems an abnormal heat-conducting properties (of an order of magnitude higher compared

	Materials	TC of PP (W/mK)	Filler content	Filler size	TC measurement method	TC of PP-composites (W/mK)
Boudenne et al. [40]	PP/Cu	0.25	35 vol%	Micron	Periodical method in [72]	2.20
Cheewawuttipong et al. [41]	PP/BN	0.25	30 vol%	Micron	THWM	2.00
Vakili et al. [47]	PP/$CaCO_3$	0.21	15 phr	Nano	GHPM	0.36
Frormann et al. [48]	PP/CNF	0.125	8 vol%	Nano	GHPM	0.181
Weidenfeller et al. [49]	PP/magnetite	0.22	44 vol%	Micron	FM	0.93
Weidenfeller et al. [50]	PP/talc	0.27	30 vol%	Micron	FM	2.50
Patti et al. [56]	PP/MWNT	0.09	5 vol%	Nano	HFMM	0.20
HeonKang'et al. [57]	PP/Al-CNT	0.20	50 vol%	Micron	FM	0.70
Chen et al. [60]	PP/PP-g-MA/f-BN	0.22	25 wt%	Micron	FM	0.60
King et al. [66]	PP/GP/CB/CNT	0.2	53.2 vol%	Micro-GP Nano-CB Nano CNT	HFMM	5.80
Krause et al. [67]	PP/GNP/CNT	0.26	7.5 vo%	Nano-CNT Micro-GNP	FM	0.50
Ren et al. [68]	PP/GS/GP	0.28	23 wt%	Nano-GP micron GS	FM	1.72
Yao et al. [69]	PP/Al_2O_3/$BaTiO_3$	0.17	50 vol%.	Sub-micron Al_2O_3 micron $BaTiO_3$	THWM	0.90
Cheewawuttipong et al. [70]	PP/VGCF/BN	0.23	53 wt%	Micron BN nano VGCF	THWM	2.90
Russo et al. [71]	PP/CNT/BN	0.09	11.5 vol%	Nano-CNT Micro-n BN	HFMM	0.25

Table 3.
Comparison among TC enhancement in some PP-based formulations due to the effect of different introduced nano-sized and/or micro-sized fillers into the matrix.

to the pristine PP) was recorded. Cheewawuttipong et al. [70] added BN and vapor-grown carbon fiber (VGCF) into the PP resin. They found that by increasing the content, the distance between fillers could be reduced and the development of conductive structure was attained. VGCF/BN hybrids possessed a better heat transport behaviour than of composites incorporating BN single size, since VGCF contributed to generate a conductive bridge by dispersing between BN and PP.

Finally, the development of carbon-based thermally conductive composites with low electrical conductivity was actualized by PP-based ternary formulations, combining CNTs (a thermal and electrical conductive filler) with additional thermally conductive, but electrically insulating, particles (ZnO,$CaCO_3$, BN, and Talc) having different sizes and shapes [71]. Results showed that, in ternary formulations, an increase of TC was always verified for all kinds of secondary particles, in particular through the union of CNTs with BN. Significant reduction of electrical conductivity was achieved, despite the presence of CNTs, with the addition of smaller secondary species (BN and ZnO), while a further increment of the same feature was obtained with larger ones ($CaCO_3$ and Talc).

Table 3 shows some values of TC reached in the PP-based composites, according to the filler type, its size and added content.

An effective comparison of reported TC values has not been possible due to differences in the filler loadings in each study. Yet, higher TC measurements were verified in the presence of microparticles, in particular with layered shape (talc); the same results have never been reached by adding nanoparticles. The highest improvement of heat transport in PP was recorded with the introduction of two combined carbonaceous nano- and micro-particles.

4. Conclusion

This review was devoted to summarizing the main literature studies about TC of polypropylene and the recent developments of heat transport ability in PP based compounds.

The TC of polypropylene has been measured in the range approximately between 0.1 and 0.2 W/mK, but different parameters as polymer cristallinity, chain structure and orientation, processing conditions and methods, temperature and pressure have played a positive or negative role on its thermal transport behaviour. In details, for the examined semi-crystalline polymer, starting from room temperature and going up to melting point, an opposite trend of TC have been reported. The TC remained almost the same as the temperature increase until it decreased during melting due to a possible breakup of crystalline regions. On the contrary, opposite results showed a strong increase of TC really at melting point. Finally, below 100 K, a growing tendency of TC against temperature has been verified. The effect of acting pressure on TC of polypropylene seemed to be always positive, probably for the induced stress in a longitudinal direction that led to anisotropy of the tested feature. For the same reason, processes like extrusion, injection molding and foaming, by causing an orientation of the polymeric molecular chains, could determine an increase of TC in the same direction of the inferred one. In the molten state, the TC of polypropylene appeared to be a complex function of molecular weight and chain branching; it continued to increase by pressure but resulted almost unaffected by temperature.

The devices, adopted for measuring the TC of PP and of its composites have been prevalently the Guarded Hot Plate Method and the Heat Flow Meter Method, based on a steady state approach, or the Flash Method and the Transient Hot Wire Method, based on transient approach.

Different efforts have been spent in literature in the improvement the heat conduction of PP, by the addition of inorganic fillers (metallics, carbon-based, ceramics and minerals) in micro- or nano-size, one-, two- or three-dimensional, with a higher thermal transmission compared to the pristine resin. By increasing filler loading, positive but not always satisfactory increases of TC in the respective compounds have been achieved. The size and shape of particles, their orientation and distribution in the polymer, the interfacial interaction between filler and matrix and between filler and filler, have been identified as crucial aspects in the optimization of final heat transport in the polymeric composites. All these factors contributed to realize effective thermally conductive pathways in the composites, actualized among particles with an advanced dispersion, good interfacial adhesion with the pristine material (lower interfacial resistance) and the proper contact among themselves (lower contact resistance).

The filler functionalization (i.e., the introduction of functional group on filler surface) and the addition of compatibilizer in polymer/particles mixtures, have been considered a useful approach for developing the compatibility between the two phases, and consequently for improving the dispersion and the interfacial interaction. Another approach has been the combination of two or more fillers, having different size and shape, to optimize the filler packing and their distribution in the matrix, so to realize and support an effective thermal conductive network.

By comparing data on PP-based compounds, despite the difference in filler loading, the greater efficiency in improving the TC of the matrix seemed to be realized in the case of combined micro- and nano-sized carbonaceous particles in the resin.

Author details

Antonella Patti[1]* and Domenico Acierno[2]

1 Department of Chemical, Materials and Industrial Engineering, University of Naples-Federico II, Naples, Italy

2 Centro Regionale di Competenza Tecnologie, Naples, Italy

*Address all correspondence to: antonlla.patti@unina.it

References

[1] Maddah HA. Polypropylene as a promising plastic: A Review. American Journal of Polymer Science. 2016;**6**:1-11. DOI: 10.5923/j.ajps.20160601.01

[2] Maier C, Calafut T. Polypropylene, The Definitive User's Guide and Databook. 1st ed. Norwich: Plastic Design Library; 1998. 452 p. ISBN 9780815518716

[3] Karian HG. Handbook of Polypropylene and Polypropylene Composites, Revised and Expanded. 2nd ed. New York: Marcel Dekker; 2003. 576 p. ISBN 9780203911808

[4] Tripathi D. Practical Guide to Polypropylene. Shrewsbury: Rapra Technology LTD.; 2002. 104 p. ISBN 9781859572825

[5] Gurses A. Introduction to Polymer-Clay Nanocomposites. 1st ed. New York: Pan Stanford; 2016. 360 p. ISBN 9789814613033

[6] Chen X, Su Y, Reay D, Riffat S. Recent research developments in polymer heat exchangers–A review. Renewable and Sustainable Energy Reviews. 2016;**60**:1367-1386. DOI: 10.1016/j.rser.2016.03.024

[7] Glade H, Moses D, Orth T. Polymer composite heat exchangers. In: Bart HJ, Scholl S, editors. Innovative Heat Exchangers. Cham: Springer; 2017. 394 p. DOI: 10.1007/978-3-319-71641-1_2

[8] Huang X, Jiang P, Tanaka T. A review of dielectric polymer composites with high thermal conductivity. IEEE Electrical Insulation Magazine. 2011;**27**:8-16. DOI: 10.1109/MEI.2011.5954064

[9] Mallik S, Ekere N, Best C, Bhatti R. Investigation of thermal management materials for automotive electronic control units. Applied Thermal Engineering. 2010;**31**:355-362. DOI: 10.1016/j.applthermaleng.2010.09.023

[10] Shojaeefard MH, Azarikhah P, Qasemian A. Experimental investigation of thermal balance and valve cover heat transfer in a small internal combustion engine. International Journal of Automotive Engineering. 2017;**7**: 2423-2434. DOI: 10.22068/ijae.7.2.2423

[11] dos Santos WN, De Sousa JA, Gregorio R. Thermal conductivity behaviour of polymers around glass transition and crystalline melting temperatures. Polymer Testing. 2013;**32**:987-994. DOI: 10.1016/j.polymertesting.2013.05.007

[12] Chen H, Ginzburg VV, Yang J, Yang Y, Liu W, Huang Y, et al. Thermal conductivity of polymer-based composites: Fundamentals and applications. Progress in Polymer Science. 2016;**59**:41-85. DOI: 10.1016/j.progpolymsci.2016.03.001

[13] Burgera N, Laachachi A, Ferriol M, Lutz M, Toniazzo V, Ruch D. Review of thermal conductivity in composites: Mechanisms, parameters and theory. Progress in Polymer Science. 2016;**61**:1-28. DOI: 10.1016/j.progpolymsci.2016.05.001

[14] Han Z, Fina A. Thermal conductivity of carbon nanotubes and their polymer nanocomposites: A review. Progress in Polymer Science. 2011;**36**:914-944. DOI: 10.1016/j.progpolymsci.2010.11.004

[15] Bird RB, Stewart WE, Lightfoot EN. Transport Phenomena. 2nd ed. New York: John Wiley & Sons, Inc; 2002. 914 p. ISBN 0-471-41077-2

[16] Anderson DR. Thermal conductivity of polymers. Chemical Reviews. 1966;**66**:677-690. DOI: 10.1021/cr60244a004

[17] Ebadi-Dehaghani H, Nazempour M. Thermal conductivity of nanoparticles filled polymers. In: Hashim A, editor. Smart Nanoparticles Technology. London: IntechOpen; 2012. 588 p. DOI:10.5772/3384

[18] Yükse N. The review of some commonly used methods and techniques to measure the thermal conductivity of insulation materials. In: Almusaed A, Almssad A, editors. Insulation Materials in Context of Sustainability. London: IntechOpen; 2016. DOI: 10.5772/64157

[19] Chapter 16. Determination of Thermal Conductivity [Internet]. 2016. Available from: http://tpm.fsv.cvut.cz/student/documents/files/BUM1/Chapter16.pdf

[20] Gaal PS, Thermitus M-A, Stroe DE. Thermal conductivity measurements using the flash method. Journal of Thermal Analysis and Calorimetry. 2004;**78**:185-189. DOI: 10.1023/B:JTAN.0000042166.64587.33

[21] Rode HM. A transient hot wire thermal conductivity apparatus for fluids. Journal of Research of the National Bureau of Standards. 1981;**86**:457-480. DOI: 10.6028/jres.086.020

[22] Jen C, Fair PG. Determination of thermal conductivity by differential scanning calorimetry. Thermochimica Acta. 1979;**34**:267-273. DOI: 10.1016/0040-6031(79)87116-6

[23] Hu M, Yu D, Wei J. Thermal conductivity determination of small polymer samples by differential scanning calorimetry. Polymer Testing. 2007;**26**:333-337. DOI: 10.1016/j.polymertesting.2006.11.003

[24] Merzlyakov M, Schick C. Thermal conductivity from dynamic response of DSC. Thermochimica Acta. 2001;**377**:183-191. DOI: 10.1016/S0040-6031(01)00553-6

[25] Bashirov AB, Manukyan AM. Thermal conductivity of polymers at various temperatures and pressures. Polymer Mechanics. 1974;**10**:484-486. DOI: 10.1007/BF00865620

[26] Osswald T, Hernandez-Ortiz JP. Polymer Processing: Modeling and Simulation. Munich: HANSER; 2006. 606 p. ISBNs 978-1-56990-398-8

[27] Choy CL, Greig D. The low temperature thermal conductivity of isotropic and oriented polymers. Journal of Physics C: Solid State Physics. 1977;**10**:169-179. DOI: 10.1088/0022-3719/10/2/005

[28] Barucci M, Gottardi E, Olivieri E, Pasca E, Risegari E, Ventura G. Low-temperature thermal properties of polypropylene. Cryogenics. 2002;**42**:551-555. DOI: 10.1016/S0011-2275(02)00076-0

[29] Dawson A, Rides M, Nottay J. The effect of pressure on the thermal conductivity of polymer melts. Polymer Testing. 2006;**25**:268-275. DOI: 10.1016/j.polymertesting.2005.10.001

[30] Andersson P, Sundqvist B. Pressure dependence of the thermal conductivity, thermal diffusivity, and specific heat of some polymers. Journal of Polymer Science: Polymer Physics Edition. 1975;**13**:243-251. DOI: 10.1002/pol.1975.180130202

[31] Eiermann K. Thermal conductivity of high polymers. Journal of Polymer Science, Polymer Symposia. 1964;**6**:157-165. DOI: 10.1002/polc.5070060118

[32] Dai SC, Tanner RI. Anisotropic thermal conductivity in sheared polypropylene. Rheologica Acta. 2006;**45**:228-238. DOI: 10.1007/s00397-005-0012-z

[33] Antunes M, Realinho V, Velasco JI, Solórzano E, Rodríguez-Pérez M-A, de Saja JA. Thermal conductivity

anisotropy in polypropylene foams prepared by supercritical CO_2 dissolution. Materials Chemistry and Physics. 2012;**136**:268-276. DOI: 10.1016/j.matchemphys.2012.07.001

[34] Ramsey JC, Fricke AL, Caskey JA. Thermal conductivity of polymer melts. Journal of Applied Polymer Science. 1973;**17**:1597-1605. DOI: 10.1002/app.1973.070170522

[35] Goswami Y. Principles of Solar Engineering. 3rd ed. Abingdon: Taylor & Francis Group; 2000. 706 p. ISBN 978-1-4665-6378-0

[36] Maier C, Calafut T. Polypropylene: The Definitive User's Guide and Databook. Norwich: William Andrew; 1998. 452 p. ISBN 9780815518716

[37] Guo B, Song Ghalambor A, Chacko J. Offshore Pipelines. Amsterdam: Elsevier; 2005 304 p. ISBN 9780080456904

[38] Birley AW. Plastic Materials: Properties and Applications. Berlin: Springer Science & Business Media; 2012. 198 p. ISBN 9789401176149

[39] Hussain ARJ, Alahyari AA, Eastman SA, Thibaud-Erkey C, Johnston S, Sobkowicz MJ. Review of polymers for heat exchanger applications: Factors concerning thermal conductivity. Applied Thermal Engineering. 2017;**113**:1118-1127. DOI: 10.1016/j.applthermaleng.2016.11.041

[40] Boudenne A, Ibos L, Fois M, Majeste JC, Gehin E. Electrical and thermal behavior of polypropylene filled with copper particles. Composites: Part A. 2005;**36**:1545-1554. DOI: 10.1016/j.compositesa.2005.02.005; 10.1016/j.compositesa.2005.02.005

[41] Cheewawuttipong W, Fuoka D, Tanoue S, Uematsu H, Iemoto Y. Thermal and mechanical properties of polypropylene/boron nitride composites. Energy Procedia. 2013;**34**:808-817. DOI: 10.1016/j.egypro.2013.06.817

[42] Glouannec P, Chauvelon P, Feller J-F, Noel H, Ploteau JP. Current passage tubes in conductive polymer composite for fluid heating. Energy Conversion and Management. 2008;**49**:493-505. DOI: 10.1016/j.enconman.2007.08.013

[43] Qin Y, Li B, Wang S. Experimental investigation of a novel polymeric heat exchanger using modified polypropylene hollow fibers. Industrial and Engineering Chemistry Research. 2012;**51**:882-890. DOI: 10.1021/ie202075a

[44] Kashiwagi T, Grulke E, Hilding J, Groth K, Harris R, Butler K, et al. Thermal and flammability properties of polypropylene/carbon nanotube nanocomposites. Polymer. 2004;**45**:4227-4239. DOI: 10.1016/j.polymer.2004.03.088

[45] Roussel MD, Guy AR, Shaw LG, Cara JE. The use of calcium carbonate in polyolefins offers significant improvement in productivity. Targets. 2005;**300**:350-355

[46] Markarian J. Mineral Modifiers Take on New Roles. Plastics, Additives and Compounding [Internet]. 2004. Available from: https://www.pharosproject.net/uploads/files/sources/1828/1359649262.pdf

[47] Vakili MH, Ebadi-Dehaghani H, Haghshenas_Fard M. Crystallization and thermal conductivity of $CaCO_3$ nanoparticle filled Polypropylene. Journal of Macromolecular Science Part B. 2011;**50**:1637-1645. DOI: 10.1080/00222348.2010.543033

[48] Frormann L, Iqbal A, Abdullah SA. Thermo-viscoelastic behavior of PCNF-filled Polypropylene nanocomposites. Journal of Applied Polymer Science. 2008;**107**:2695-2703. DOI: 10.1002/app.27349

[49] Weidenfeller B, Höfer M, Schilling F. Thermal and electrical properties of magnetite filled polymers. Composites Part A: Applied Science and Manufacturing. 2002;**33**:1041-1053. DOI: 10.1016/S1359-835X(02)00085-4

[50] Weidenfeller B, Höfer M, Schilling F. Thermal conductivity, thermal diffusivity, and specific heat capacity of particle filled polypropylene. Composites Part A: Applied Science and Manufacturing. 2004;**35**:423-429. DOI: 10.1016/j.compositesa.2003.11.005

[51] Andrews R, Weisenberger MC. Carbon nanotube polymer composites. Current Opinion in Solid State and Materials Science. 2004;**8**:31-37. DOI: 10.1016/j.cossms.2003.10.006

[52] Pollack GL. Kapitza. Resistance. Review of Modern Physics. 1969;**41**: 48-81. DOI: 10.1103/RevModPhys.41.48

[53] Swartz ET, Pohl RO. Thermal boundary resistance. Review of Modern Physics. 1989;**61**:605-668. DOI: 10.1103/RevModPhys.61.605

[54] Vidal-Sallé E, Marchand A-S. Modelling of the friction thermo-mechanical coupling at the workpiece-tool interface during bulk forming processes. Tribology Series. 2003;**43**:349-356. DOI: 10.1016/S0167-8922(03)80062-7

[55] Hussain F, Hojjati M, Okamoto M, Goiga RE. Review article: Polymer-matrix nanocomposites, processing, manufacturing, and application: An overview. Journal of Composite Materials. 2006;**40**:1511-1575. DOI: 10.1177/0021998306067321

[56] Patti A, Russo P, Acierno D, Acierno S. The effect of filler functionalization on dispersion and thermal conductivity of polypropylene/multi wall carbon nanotubes composites. Composites Part B Engineering. 2016;**94**:350-359. DOI: 10.1016/j.compositesb.2016.03.072

[57] HeonKang CH, Yoon KH, Park Y-B, Lee DY, Jeong SS. Properties of polypropylene composites containing aluminum/multi-walled carbon nanotubes. Composites Part A: Applied Science and Manufacturing. 2010;**41**:919-926. DOI: 10.1016/j.compositesa.2010.03.011

[58] Muratov DS, Kuznetsov DV, Il'inykh IA, Burmistrov IN, Mazov IN. Thermal conductivity of polypropylene composites filled with silane-modified hexagonal BN. Composites Science and Technology. 2015;**111**:40-43. DOI: 10.1016/j.compscitech.2015.03.003

[59] Ezat GS, Kelly AL, Mitchell SC, Youseffi M, Coates PD. Effect of maleic anhydride grafted Polypropylene Compatibilizer on the morphology and properties of Polypropylene/multiwalled carbon nanotube. Composites. 2012;**33**:1376-1386. DOI: 10.1002/pc.22264

[60] Chen L, Xu H-F, He S-J, Yi-Hang D, Yu N-J, Du X-Z, et al. Thermal conductivity performance of Polypropylene composites filled with Polydopamine-functionalized hexagonal boron nitride. PLoS One. 2017;**12**:e0170523. DOI: 10.1371/journal.pone.0170523

[61] Szentes A, Varga CS, Horváth G, Bartha L, Kónya Z, Haspel H, et al. Electrical resistivity and thermal properties of compatibilized multi-walled carbon nanotube/polypropylene composites. eXPRESS Polymer Letters. 2012;**6**:494-502. DOI: 10.3144/expresspolymlett.2012.52

[62] Díez-Pascual AM, Naffakh M, González-Domínguez JM, Ansón AM, Martínez-Rubi Y, Martínez MT, et al. High performance PEEK/carbon nanotube composites compatibilized with polysulfones-II. Mechanical and electrical properties. Carbon. 2010;**48**:3500-3511. DOI: 10.1016/j.carbon.2010.05.050

[63] Patti A, Acierno D, Russo P. Influence of filler dispersion and interfacial resistance on thermal conductivity of Polypropylene/carbon nanotubes systems. AIP Conference Proceedings. 2017;**1914**:030014. DOI: 10.1063/1.5016701

[64] Ashby MF, Brechet YJM. Designing hybrid materials. Acta Materialia. 2003;**51**:5801-5821. DOI: 10.1016/S1359-6454(03)00441-5

[65] Szeluga U. Kumanek., Trzebicka B. Synergy in hybrid polymer/nanocarbon composites. A review. Composites: Part A. 2015;**73**:204-231

[66] King JA, Johnson BA, Via MD, Ciarkowski CJ. Effects of Carbon Fillers in Thermally Conductive Polypropylene Based Resins. Polymer Composites. 2010;**31**:497-506. DOI: 10.1002/pc.20830

[67] Krause B, Pötschke P. Electrical and TC of polypropylene filled with combinations of carbon fillers. AIP Conference Proceedings. 2016;**1779**:040003. DOI: 10.1063/1.4965494

[68] Ren Y, Zhang Y, Fang H, Ding T, Li J, Bai S-L. Simultaneous enhancement on thermal and mechanical properties of polypropylene composites filled with graphite platelets and graphene sheets. Composites: Part A. 2018;**112**:57-63. DOI: 10.1016/j.compositesa.2018.05.017

[69] Yao J, Hu L, Zhou M, You F, Jiang X, Gao L, et al. Synergistic enhancement of thermal conductivity and dielectric properties in Al_2O_3/$BaTiO_3$/PP. composites. Materials. 2018;**11**:1536. DOI: 10.3390/ma11091536

[70] Cheewawuttipong W, Tanoue S, Uematsu H, Iemoto Y. Thermal conductivity of polypropylene composites with hybrid fillers of boron nitride and vapor-grown carbon fiber. Polymer Composites. 2016;**37**:936-942. DOI: 10.1002/pc.23252

[71] Russo P, Patti A, Petrarca C, Acierno S. Thermal conductivity and dielectric properties of polypropylene-based hybrid compounds containing multiwalled carbon nanotubes. Journal of Applied Polymer Science. 2018;**135**:46470-46479. DOI: 10.1002/app.46470

[72] Boudenne A, Ibos L, Gehin E, Candau Y. A simultaneous characterization of thermal conductivity and diffusivity of polymer materials by a periodic method. Journal of Applied Physics. 2004;**37**:132-139. DOI: 10.1088/0022-3727/37/1/022

Chapter 4

Polypropylene Blends and Composite: Processing-Morphology-Performance Relationship of Injected Pieces

Maria Alejandra Costantino, Caren Rosales and Valeria Pettarin

Abstract

Polypropylene (PP) is a low-cost plastic commodity, which currently is in a transition zone between massive use and engineering applications due mainly to its limited mechanical properties, such as low tensile and impact resistance. That is the reason why PP is usually modified with additives and particles to improve its mechanical and thermal performance and thus meet the requirements demanded by engineering applications. Besides, PP composites are suitable materials to be processed by a simple, fast, automatic, and massive technique such as injection molding. This makes PP composites attractive for several applications. However, it is important to keep in mind that PP composites' performance depends not only on their intrinsic properties but also on processing conditions. This chapter will summarize the relationship between processing and performance of several PP composite—micro, nano, and hybrid—injected parts with the aim of generating a bridge between technologic knowledge and scientist knowledge.

Keywords: polypropylene, injection molding, microcomposites, nanocomposites, hybrid composites

1. Introduction

Polypropylene (PP) is a plastic commodity, which currently is in a transition zone between massive use and engineer applications due mainly to its limited mechanical properties, such as low tensile and impact resistance. Moreover, several years ago the replacement of conventional materials with lighter ones has attracted the attention of many industries, especially the automotive ones. Replacement of traditional materials is achieved with the development of new composite materials, which meet both the desired properties—mechanical, thermal, esthetic—and a low weight, i.e., high relative properties. These are the main reasons why PP is usually modified with additives and particles to improve its mechanical and thermal performance and thus meet the requirements demanded by engineer applications. Originally, PPs were modified with fillers as talc only to reduce their costs, but currently the purpose is focused on improving properties such as rigidity, strength, toughness, dimensional stability, and even esthetics of PP parts [1]. To achieve this purpose, several particles have been used, such as glass fibers (GF), nanoclay (NC), carbon nanotubes (CNT), and rubber. In a further step, hybrid materials formulated by two or more of these components have been proposed.

From an engineering point of view, mixing a polymer matrix with a particle is an effective low-cost way to achieve the required properties when parts are produced by injection molding. However, it is important to understand the way in which particles and processing affect the structure and properties of processed parts. It is important to keep in mind that PP composites' performance depends not only on their intrinsic properties but also on processing conditions. PP is also strongly sensitive to defects produced during manufacturing processes such as injection molding, which deteriorate and decrease a lifetime of composite parts [2]. These defects are even more pronounced in the case of composites. In recent years, a number of texts regarding properties of injection-molded reinforced polypropylenes have been published [3–5]. However, because of the continuing developments of PP composites, the achievable property values are continuing to improve. In structural and semi-structural applications, particularly, in addition to high stiffness and mechanical strength, adequate fracture toughness is often required. In order to optimize these properties, the knowledge of the relationship between morphology and deformation behavior seems to be essential. The understanding of the fracture, micro-deformation, and mechanics of failure of composites is therefore crucial for engineers. This chapter will summarize the relationship between processing and performance of several PP composite—micro, nano, and hybrid—injected parts aiming to generate a bridge between technologic knowledge and scientist knowledge.

1.1 Injection molding process

Injection molding of thermoplastic polymers is a repetitive process in which a molten polymer is forced to go through a mold cavity where it is held under low pressure until it solidifies and it is finally ejected. A scheme of injection molding machine can be observed in **Figure 1**.

First stage of injection cycle begins with the molten polymer filling the cavity mold which is closed (filling stage). During filling stage, the screw doesn't rotate but acts as a dashpot which drives the molten material into the mold. At the end of the filling stage, a lower pressure is held by the feeding system allowing a small amount of additional material to enter the mold cavity to compensate the volumetric contraction of the injected part (holding stage). Holding pressure eventually decreases to zero; this defines the beginning of the third stage known as "cooling stage." In this stage, the molten material solidifies inside the cavity, the

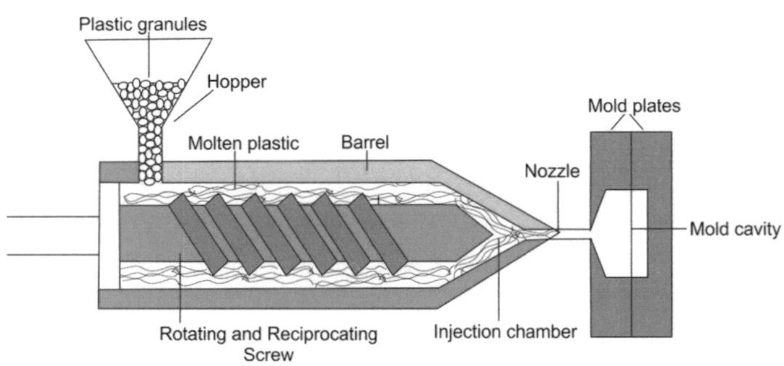

Figure 1.
Injection molding machine scheme.

mold opens, and finally the piece is ejected. The mold then closes again to start a new cycle [6].

1.2 Injection molding common defects

Although the main advantage of injection molding process is to manufacture complex parts in a single, fast, and automatic operation, there still are some processing inherent defects such as flow and weld lines which may deteriorate the mechanical performance and appearance of final injected parts. Weld lines are the result of the convergence of several flow fronts during filling stage. The origin of these flow fronts may be due to several reasons: inserts inside the mold cavity, thickness differences in the piece, or the presence of two or more injection points [7]. Weld lines are usually V-shaped as it can be seen in **Figure 2**.

Cross section of the welding plane shows two different zones within the weld line with some particular characteristics: a V-shaped zone where there is almost no molecular diffusion and unfavorable orientation and a central zone with a better molecular diffusion (see **Figure 2**). Weld line is a weak zone from a mechanical point of view and uses to present visual defects too [8, 9]. Weld line performance is determined by material nature, part complexity, and processing variables.

Another common and important injection molding defect is warping. Warping is a macro-geometric deformation of injected pieces which remains after cooling. The main causes of warping are differential contraction between different parts of pieces and released residual stresses formed during cooling stage. These deformations are mainly due to confinement of pieces in the mold cavity, orientation, crystallization, or cooling differential.

1.3 Injected polypropylene

PP performance—mechanical, thermal, and electrical—depends mainly on its morphology and crystallinity [10]; and processing affects both morphology and crystallinity of polymers. In the case of injection molding, the molten polymer is subjected to thermomechanical complex conditions characterized by high cooling speeds and stress fields. These conditions change along the flow path and mold thickness, i.e., polymer pieces present an intrinsic heterogeneous microstructure, characterized by a gradual and hierarchical variation of morphology, which evolves through the spatial domain of the piece. Injected PP particularly develops a "skin-core"

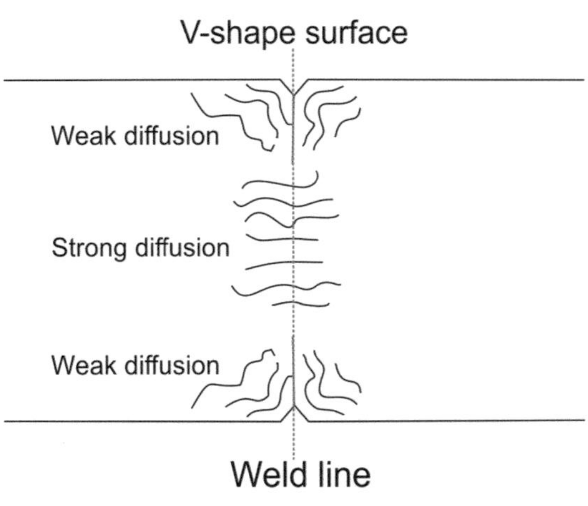

Figure 2.
Cross section of a weld line zone.

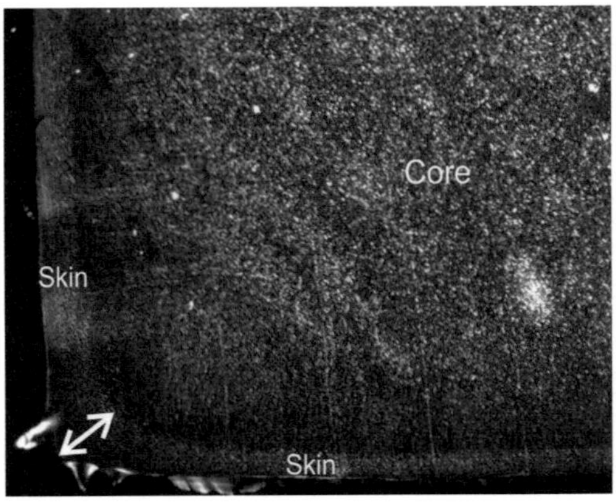

Figure 3.
Skin-core microstructure seen by polarized optical microscopy.

microstructure, which can be seen by polarized optical microscopy (PLM), as in **Figure 3**.

The number of observed "layers" in the microstructure depends on the resolution of the experimental technique used. A simple analysis considers a three-layer model (two external skins and an inner core) [11–13], but other layers may be also observed (two external skins, two sub-skin regions, two shear zones, and an inner core). The intrinsic molecular nature of the polymer together with this layer morphology determines the mechanical performance of injected PP parts.

Besides, adding a second component—particles or additives—into a PP matrix may also change its crystalline structure, i.e., may produce changes in injected piece performance.

Through this chapter, the relationship between processing and performance is reviewed for injected PP composites. The combined effect of the molding process and the fillers on the properties of the polymer composites is reviewed. Also, the effects of the occurrence of inhomogeneities, such as weld lines or flow lines in microstructure and therefore in performance, are summarized.

2. Injected PP microcomposites

The first attempt to obtain a good composite is to add a microparticle to the polymer matrix. Among PP microcomposites, fiber-reinforced plastics are a popular type of composites used in many engineering applications mainly because of their excellent capability to form complex shapes. These fibers—glass or carbon, stiff and elastic—generally increase both stiffness and strength of PP matrixes. Even though injection molding is currently the most used technique to process this kind of composites, there are some issues that directly link the processing with a nonuniform orientation of fibers and their breakage. In fact, there is a strong heterogeneity of the microstructure in terms of fiber orientation of injected parts: short-glass fibers use to tend parallel to the injection flow direction in the skin zones and highly angled with respect to flow direction in the core layer. Fiber orientation depends also on location along the piece (e.g., distance from injection points). There is a strong dependence of the macroscopic mechanical behavior on fiber orientation. In fact, when the average angle of fiber orientation varies in only a few degrees with respect to loading direction—corresponding with a change in the average value of the component of fiber orientation matrix with respect to loading direction of only a few hundredths—the composite tensile strength varies by approximately 7.5% [14]. In addition to orientation, fiber length also dominates the tensile strength of injection-molded composites. In general, fiber breakage results in a decrease in tensile strength, so it is important to know how injection molding affects fiber breakage. An increase in shear (injection velocity, shear components, etc.) may cause an enhanced fiber and matrix orientation which would lead to higher tensile performance along the flow direction. However, it also causes a remarkable fiber breakage [15]. Experimental results indicate that an increase in injection velocity results in a decrease in ultimate tensile stress, due to the high fiber breakage. This effect is partly attenuated at low mold temperature, due to an increase of fiber orientation [3]. In case of complex parts that contain weld lines, the situation is even more complicated. The fibers are nonuniformly distributed in the regions around the defects, and there is also a distribution of glass fiber densities. All these features modify fracture behavior of injected pieces changing failure patterns, with crack pathways that follow stress concentrators developed during processing [4].

As it was stated before, not only all modifications are done to obtain a nanocomposite with improved mechanical performance, but also esthetic features are searched for some special applications. Composites of thermoplastic polymers with metallic fillers are an important group of engineering materials with a wide range of properties including electric and thermal conduction, high mechanical properties, and improved esthetic quality. Currently, metallic looking plastics replace metals by plastic in many applications, trying to achieve the quality and prestige of metals and adding value to products [16]. It is possible to obtain a metallic looking plastic part by adding metallic pigments. In this way, it is possible to eliminate post-processing operations such as painting. Metallic pigments have different shapes and sizes. Particles with a flake shape promote the reflected light in a specular way increasing the metallic appearance of part surfaces [17]. In spite of the injection defects—as weld or flow lines—being known to affect pieces of esthetics, this could be improved by adjusting processing conditions [18, 19]. Melt temperature is one of the processing parameter that more influences esthetic of injected parts: higher melt temperatures decrease shrinkage and make weld lines wider and more diffuse [20]. In case of PP/aluminum composites, the presence of aluminum that increases thermal conductivity plus the inherent temperature gradient and shear stresses of the injection molding induces β-polymorph formation. This effect also depends on processing conditions; a higher melt temperature induces a higher β-phase content. At the same time, the mechanical performance of parts shows to be dependent on PP morphology, i.e., processing conditions. Quasi-static fracture performance also depends on the location of the samples. At weld line zone, PP/aluminum composite failed in a brittle way following the weld line. Fracture toughness of both PP and PP/aluminum is similar, indicating that weld lines are a predominant weak defect inside the injected parts. Away from weld lines, PP and PP/aluminum show a

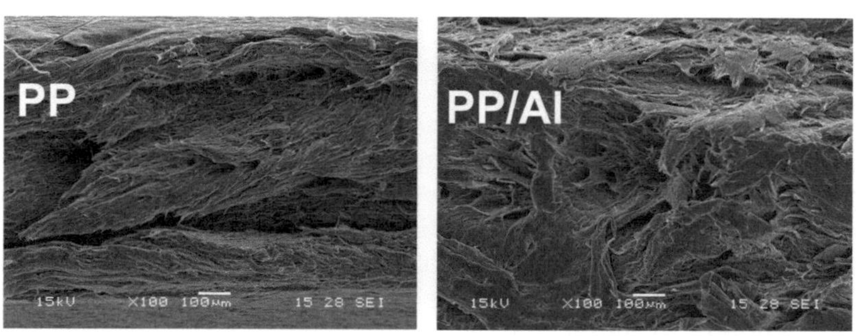

Figure 4.
SEM pictures of PP and PP/Al fracture surfaces.

similar fracture behavior characterized by a nonlinearity of stress vs. strain curves with a crack stable propagation and large plastic deformation. However, differences in the propagation mode between PP and PP/aluminum parts were found. In fact, specific plastic work w_p (specific energy absorption per unit volume) is equal to 4.51 MJ/m^3 for neat PP and 16.01 MJ/m^3 for the PP of PP/Al composite. These values indicate that much more energy is involved in the propagation of a crack in the PP of PP/Al samples than in the PP of neat PP samples. The occurrence of β-phase in the composite promotes matrix fibrillation and makes PP of PP/aluminum parts to consume more energy before break than pure PP injected parts (**Figure 4**).

3. Injected PP nanocomposites

In the last three decades, a large interest in nanocomposites was seen in both academic and industry fields [21], due to their potential improvement of properties with a low content of the second phase. In the case of nanocomposites, nanofiller dispersion and orientation are very determinant of mechanical and thermal properties. In theory, only well-dispersed and exfoliated (nanoclays) nanoparticles could lead to an effective improvement of composite performance. Most of commercial nanoparticles, such as nanoclay, are hydrophilic and have weak adhesion or interaction with a hydrophobic matrix as PP, leading to a nanocomposite with a poor dispersion. As a solution to this problem, some producers recommend using masterbatches (MBs), which include all compatibilizers needed to promote nanoclay dispersion and have the additional advantage of being easy to process and compatible with standard processing as injection molding. In fact, some authors have reported nanocomposite preparation by using MB [22–25]. However, these studies indicate that nanoparticles were not exfoliated but intercalated. The influence of the flow pattern during injection molding on the fracture and impact properties of complex injected PP/clay nanocomposites has been studied [22, 23]. Nanoparticles were mostly intercalated—even though they were chemically modified and compatibilized—and both fracture parameters (KIC and G) and impact toughness were determined by molecular and nanoparticle orientation induced by the flow pattern. Toughness mechanisms—as particle delamination or separation—were active only in certain loading directions. It was stated before that nanoclay delamination or splitting is an effective toughness mechanism in nanocomposites [26]. In PP, craze-like bands are one of the main mechanisms responsible for matrix energy absorption during deformation. To activate this mechanism, free surfaces are necessary for craze bands to initiate and nanoclay delamination produces those surfaces. However, these crazes can initiate only at the pole of clay particles, i.e., only particles oriented at 45° or more to load direction can induce multiple crazing in tensile-loaded samples and subsequently act as a toughening mechanism. Besides, weld and flow lines produced during filling acted as defects in the presence of nanoparticles [22, 23]. For PP/nanoclay composites under tensile conditions, the amount of absorbed energy was lower at the weld line than away from it and in the flow direction. This is a clear example of how injection molding flow pattern affects the piece performance of injected nanocomposites.

As nanofiller dispersion and exfoliation are crucial, a great effort has been made to improve them by adding additives. However, it is also possible to improve dispersion and exfoliation by changing their processing characteristics. An example of this is shear-controlled orientation in injection molding (SCORIM), which is a not conventional injection molding technique based on a shear-controlled application to the molten polymer during holding stage. SCORIM involves the use of a conventional injection molding machine with a special device with two

Skin orientation (A_{110} index)	Percentage of crystallinity (%)	J integral (N/mm)
0.15	31	94.6
0.165	37	83.8
0.17	40	73
0.189	37.5	40.6
0.19	41	51.4
0.195	39	29.8
0.20	42	8.6

Table 1.
Orientation, percentage of crystallinity, and fracture energy values [34].

pistons that generate the shear stresses. It was reported that SCORIM improves the performance of injected parts by controlling their morphology [27, 28]. Significant improvements were found in both stiffness and tensile resistance, molecular and filler orientation, dimensional tolerance, esthetic appearance, and weld line elimination in PP [29] and in its nanocomposites [30–32]. It was demonstrated that SCORIM changes the morphology of PP nanocomposites, not only in terms of molecular and nanoclay orientation but also in crystal phases present in PP matrix: the shear stresses driven by SCORIM process induce the formation of γ phase in PP nanocomposites [24, 33]. SCORIM induces a thicker skin in nanocomposites, i.e., a larger proportion of orientated molecules and clay particles, which favors the sliding of macromolecules, improving the deformation capability. Meanwhile, γ polymorph induces a larger-scale plastic deformation compared with the common α phase. γ phase promotes tearing of PP ligaments leading to fibrillation which is a toughness mechanism [34]. All these morphology features—better molecular and particle orientation and γ polymorph presence—improved PP/nanoclay toughness (**Table 1**).

In case of carbon nanotubes (CNTs), it is important not only to attain a good dispersion but also to obtain an interconnected network morphology (above the percolation threshold) to lead an improvement in composites' performance. This morphology depends on nanotube orientation, dispersion, and distribution [35]. It is known that injection-molded parts have a higher percolation threshold than compressed ones due to the morphology and orientation induced by processing [36–38]. Moreover, weld lines could make particle dispersion and orientation even more complex for this kind of materials. It was reported that PP-/CNT-injected parts presented also agglomerates and an isotropic morphology induced by flow pattern during injection [39]. Also, an orientation profile of CNT trough molding thickness has been seen near to the injection point. CNT particles in the skin zone are oriented parallel to the flow front, while they tend to align transversally in the core zone. This has also been observed for fiber-reinforced polymers [40, 41]. In weld line region particularly, it was reported that agglomerates are more diffuse with a random orientation of CNT in both skin and core zones [42]. This heterogeneous orientation induces different fracture mechanisms in the pieces, weld line zone being the weakest part of injected pieces. Agglomerates act also as defects diminishing fracture energy of nanocomposites, when compared with pure PP. As a result of this particle orientation, electrical conductivity—both AC and DC—also changes along injected pieces: at the weld line region, there is an increase in conductivity values due to the more efficient conductive filler distribution. In this example, it can be clearly seen that morphology developed during injection molding is a crucial feature which

	Electrical resistivity (Ohm/cm)	e-Painting efficiency	G_C (kJ/m^2)
PP/CNT weld line	1.10^9	0.7	3.3
PP/CNT bulk	1.10^{10}	0.6	8.4
Neat PP	1.10^{14}	0.1	18.1

Table 2.
Electrical, e-painting efficiency, and fracture energy values [42].

determines piece performance. A 3D interconnected CNT network is optimal to obtain good electrical conductivity values, but it is not favorable for obtaining a good fracture performance (since it inhibits alternative toughness mechanisms to occur) (**Table 2**).

4. Injected PP hybrids

The use of two different reinforcements at the same time may expand the application field of PP composites by combining their properties. Hybrid nanocomposites made by a rigid filler and soft particles have attracted attention due to incorporation of both stiffness and higher energy absorption and elongation at break. The goal is to obtain an optimal balance between rigidity and impact resistance [43–45]. An example of these kinds of hybrid composites is a rubber/nanoclay/polypropylene nanocomposite, which may increase simultaneously stiffness and toughness of PP. A study about how injection molding flow pattern and inhomogeneities affect the morphology and performance of this hybrid nanocomposite at different locations in injected intricate moldings was reported in literature [46]. A noticeable morphological feature was found: rubber particles appear to be more elongated and oriented in flow direction in the skin of injected pieces, while they appear spherical shaped in the core. Regarding nanoclay, an orientation profile was found: there is a strong orientation of nanoparticles in flow direction in skin zone, while they are randomly distributed in the core. A scheme of these morphology features is shown in **Figure 5**. Surprisingly, there are no significant morphological differences between the zone near the injection point and the zone of the weld line. These morphological features have an important influence in mechanical performance of injected pieces. In fact, fracture features showed to be dependent on the morphology developed during processing: in the core—with spherical-shaped rubber particles and randomly oriented nanoclay—a cavitation process was seen accompanied by shear yielding; in the skin, with elongated-shaped rubber particles and strongly orientated nanoclay particles in flow direction, there were no signs of cavitation, and fracture surface was slightly rugged. It is known that size and shape of rubber particles play a key role in

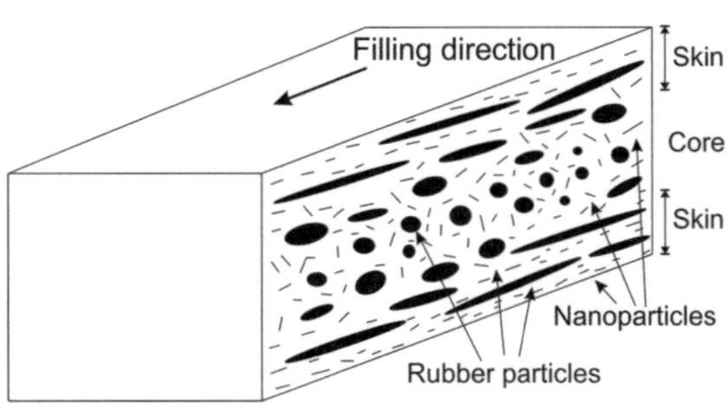

Figure 5.
Particle morphology feature scheme.

toughness mechanisms [47–50]. Spherical particles favor stress concentration which induces several absorption mechanisms—craze, shear yielding—while elongated ones are not able to produce stress fields needed to promote toughness mechanisms [50]. The non-visibility of the weld line seen previously in the morphological analysis is in concordance with fracture results: weld lines did not act as stress raisers and did not introduce an alternative crack path.

Another kind of a hybrid composite which combines both reinforcement and energy consumption promotion is composites reinforced with both glass and cellulose fibers. Kahl et al. studied the synergetic effect of those two fibers in hybrid injected PP-based composites. They found that fiber orientation depends not only on the flow pattern but also on the amount of cellulose fibers present in the hybrid composite. There is a general trend of both kinds of fibers to orientate parallel to flow direction. A higher cellulose fiber content in the hybrid composite decreases the orientation of both fibers. Besides, short fibers tend to align following flow direction much more than longer ones [51]. There are some other examples of works in which fiber interactions govern the morphology of hybrid injected pieces. Gamze et al. found that mechanical performance of injected carbon nanotubes/glass fiber PP hybrid composites depends on fiber interaction in the matrix. The simultaneous usage of carbon nanotubes and glass fibers increases the system polarity, leading to a better dispersion of carbon nanotubes, with the subsequent effect in the final performance of pieces [52].

In summary, there are several evidences indicating that not only the developed morphology during processing but also the interaction between reinforcements is crucial in the final performance of injected pieces.

5. Conclusions

Through this chapter, it has been shown that PP composites' performance depends not only on their intrinsic properties but also on processing conditions. Processing of a two- or three-phase PP-based composite induces distinct morphologies and microstructures that depend on both processing conditions and phase interaction, i.e.:

- The same composite would develop different morphologies or microstructures if processed with different conditions.

- Different composites processed with the same conditions would also develop different morphologies or microstructures.

These induced characteristics, such as crystallinity, crystalline phase, or phase morphology, will definitely affect final performance of processed pieces, including thermal, mechanical, and fracture behaviors. Moreover, if different types of reinforcements are added in a composite, it has been observed that not only the developed morphology during processing but also the interaction between reinforcements is crucial in the final performance of pieces.

All these features should be kept in mind when trying to use a composite, knowing that laboratory results should not be directly extrapolated to final processed pieces.

Author details

Maria Alejandra Costantino*, Caren Rosales and Valeria Pettarin
Institute of Materials Science and Technology (INTEMA),
University of Mar del Plata, Mar del Plata, Argentina

*Address all correspondence to: Alejandra-costantino@hotmail.com

References

[1] Karger-Kocsis J, editor. Polypropylene Structure, Blends and Composites: Volume 1 Structure and Morphology. Netherlands: Springer; 1995

[2] Wright DC. Failure of Plastics and Rubber Products: Causes, Effects and Case Studies Involving Degradation. Shawbury: Smithers Rapra Press; 2001

[3] Crema L, Sorgato M, Zanini F, Carmignato S, Lucchetta G. Experimental analysis of mechanical properties and microstructure of long glass fiber reinforced polypropylene processed by rapid heat cycle injection molding. Composites Part A: Applied Science and Manufacturing. 2018;**107**:366-373. DOI: 10.1016/j.compositesa.2018.01.019

[4] Lee J-M, Moon J-S, Shim D, Choi B-H. Effect of glass fiber distributions on the mechanical and fracture behaviors of injection-molded glass fiber-filled polypropylene with 2-Hole Tension specimens. Composites Science and Technology. 2019;**170**:190-199. DOI: 10.1016/j.compscitech.2018.11.038

[5] Li X, Gong N, Yang C, Zeng S, Fu S, Zhang K. Aluminum/polypropylene composites produced through injection molding. Journal of Materials Processing Technology. 2018;**255**:635-643. DOI: 10.1016/j.jmatprotec.2018.01.008

[6] Kennedy P, Zheng R. Flow Analysis of Injection Molds. Munich; Cincinnati: Hanser Fachbuchverlag; 2013

[7] Malloy RA. Plastic Part Design for Injection Molding: An Introduction. München (Alemania): Carl Hanser Verlag; 1994

[8] Brostow W, Corneliussen RD. Failure of Plastics. Munich; New York: Hanser Pub.: Distributed in the United States of America by Macmillan Pub.; 1986

[9] Dealy JM, Wissbrun KF. Melt Rheology and its Role in Plastics Processing: Theory and Applications. Netherlands: Springer; 1999

[10] Frontini PM, Fave A. The effect of annealing temperature on the fracture performance of isotactic polypropylene. Journal of Materials Science. 1995;**30**:2446-2454. DOI: 10.1007/BF01184599

[11] Trotignon J-P, Verdu J. Skin-core structure–fatigue behavior relationships for injection-molded parts of polypropylene. I. Influence of molecular weight and injection conditions on the morphology. Journal of Applied Polymer Science. 1987;**34**:1-18. DOI: 10.1002/app.1987.070340101

[12] Fujiyama M, Wakino T. Molecular orientation in injection-molded polypropylene copolymers with ethylene. IPP. 1992;7:97-105. DOI: 10.3139/217.920097

[13] Kantz MR, Newman HD, Stigale FH. The skin-core morphology and structure–property relationships in injection-molded polypropylene. Journal of Applied Polymer Science. 1972;**16**:1249-1260. DOI: 10.1002/app.1972.070160516

[14] Notta-Cuvier D, Nciri M, Lauro F, Delille R, Chaari F, Robache F, et al. Coupled influence of strain rate and heterogeneous fibre orientation on the mechanical behaviour of short-glass-fibre reinforced polypropylene. Mechanics of Materials. 2016;**100**:186-197. DOI: 10.1016/j.mechmat.2016.06.013

[15] Wang J, Geng C, Luo F, Liu Y, Wang K, Fu Q, et al. Shear induced fiber orientation, fiber breakage and matrix molecular orientation in long glass fiber reinforced polypropylene composites. Materials Science and

Engineering A. 2011;**528**:3169-3176. DOI: 10.1016/j.msea.2010.12.081

[16] Wheeler I. Metallic Pigments in Polymers. Shawbury, U.K.: Rapra Technology Ltd.; 1999

[17] Bunge H. Metallic looking plastics with new silver and colored aluminum pigments. In: Presented at the Annual Technical Conference—ANTEC, Atlanta, USA. 1998

[18] Kobayashi Y, Teramoto G, Kanai T. The unique flow of polypropylene at the weld line behind an obstacle in injection molding. Polymer Engineering & Science. 2011;**51**:526-531. DOI: 10.1002/pen.21841

[19] Harris RM. Coloring Technology for Plastics. Norwich, USA: William Andrew; 1999

[20] Costantino MA, Pettarin V, Pontes A, Frontini P. Mechanical performance of double gated injected metallic looking polypropylene parts. Express Polymer Letters. 2015;**9**:1040-1051. DOI: 10.3144/expresspolymlett.2015.93

[21] Jawaid M, Qaiss A e K, Bouhfid R, editors. Nanoclay Reinforced Polymer Composites: Nanocomposites and Bionanocomposites. Singapore: Springer; 2016

[22] Pettarin V, Viau G, Fasce L, Viana JC, Pontes AJ, Frontini PM, et al. Uni- and biaxial impact behavior of double-gated nanoclay-reinforced polypropylene injection moldings. Polymer Engineering & Science. 2013;**53**:724-733. DOI: 10.1002/pen.23306

[23] Pettarin V, Brun F, Viana JC, Pouzada AS, Frontini PM. Toughness distribution in complex PP/nanoclay injected mouldings. Composites Science and Technology. 2013;**74**:28-36. DOI: 10.1016/j.compscitech.2012.09.015

[24] Costantino A, Pettarin V, Viana J, Pontes A, Pouzada A, Frontini P. Morphology–performance relationship of polypropylene–nanoclay composites processed by shear controlled injection moulding. Polymer International. 2013;**62**:1589-1599. DOI: 10.1002/pi.4543

[25] Nevalainen K, Auvinen S, Orell O, Eteläaho P, Suihkonen R, Vuorinen J, et al. Characterization of melt-compounded and masterbatch-diluted polypropylene composites filled with several fillers. Polymer Composites. 2013;**34**:554-569. DOI: 10.1002/pc.22454

[26] Cotterell B, Chia JYH, Hbaieb K. Fracture mechanisms and fracture toughness in semicrystalline polymer nanocomposites. Engineering Fracture Mechanics. 2007;**74**:1054-1078. DOI: 10.1016/j.engfracmech.2006.12.023

[27] Kalay G, Bevis MJ. Processing and physical property relationships in injection-molded isotactic polypropylene. 2. Morphology and crystallinity. Journal of Polymer Science Part B: Polymer Physics. 1997;**35**: 265-291. DOI: 10.1002/(SICI)1099-0488(19970130)35:2<265:AID-POLB6>3.0.CO;2-R

[28] Kalay G, Bevis MJ. Processing and physical property relationships in injection-molded isotactic polypropylene. 1. Mechanical properties. Journal of Polymer Science Part B: Polymer Physics. 1997;**35**: 241-263. DOI: 10.1002/(SICI)1099-0488(19970130)35:2<241::AID-POLB5>3.0.CO;2-V

[29] Kikuchi A, Coulter JP, Angstadt DC. Polymer melt manipulation and in-process morphology control during molding processes: A review. Journal of Injection Molding Technology. 2002;**6**:91-106

[30] Bilewicz M, Viana JC, Cunha AM, Dobrzański LA. Morphology diversity and mechanical response of injection moulded polymer nano-composites and polymer-polymer composites.

Journal of Achievements in Materials and Manufacturing Engineering. 2006;**15**(7):159-165

[31] Bilewicz M, Viana JC, Dobrzański LA. Self reinforced polymer-polymer composites. Journal of Achievements in Materials and Manufacturing Engineering. 2007;**24**:4

[32] Dobrzański LA, Bilewicz M, Viana JC, Cunha AM. Non-conventionally obtained polymer nanocomposites with different nano-clay ratios. Journal of Achievements in Materials and Manufacturing Engineering. 2008;**31**:6

[33] Costantino A, Pettarin V, Viana J, Pontes A, Pouzada A, Frontini P. Microstructure of PP/clay nanocomposites produced by shear induced injection moulding. Procedia Materials Science. 2012;**1**:34-43. DOI: 10.1016/j.mspro.2012.06.005

[34] Yuan Q, Deshmane C, Pesacreta TC, Misra RDK. Nanoparticle effects on spherulitic structure and phase formation in polypropylene crystallized at moderately elevated pressures: The influence on fracture resistance. Materials Science and Engineering A. 2008;**480**:181-188. DOI: 10.1016/j.msea.2007.07.019

[35] Prashantha K, Soulestin J, Lacrampe MF, Krawczak P, Dupin G, Claes M. Masterbatch-based multi-walled carbon nanotube filled polypropylene nanocomposites: Assessment of rheological and mechanical properties. Composites Science and Technology. 2009;**69**:1756-1763. DOI: 10.1016/j.compscitech.2008.10.005

[36] Arjmand M, Apperley T, Okoniewski M, Sundararaj U. Comparative study of electromagnetic interference shielding properties of injection molded versus compression molded multi-walled carbon nanotube/polystyrene composites. Carbon. 2012;**50**:5126-5134. DOI: 10.1016/j.carbon.2012.06.053

[37] Van Hattum FWJ, Leer C, Viana JC, Carneiro OS, Lake ML, Bernardo CA. Conductive long fibre reinforced thermoplastics by using carbon nanofibres. Plastics, Rubber and Composites. 2006;**35**:247-252. DOI: 10.1179/174328906X146504

[38] Kalaitzidou K, Fukushima H, Drzal LT. A new compounding method for exfoliated graphite–polypropylene nanocomposites with enhanced flexural properties and lower percolation threshold. Composites Science and Technology. 2007;**67**:2045-2051. DOI: 10.1016/j.compscitech.2006.11.014

[39] Abbasi S, Carreau PJ, Derdouri A. Flow induced orientation of multiwalled carbon nanotubes in polycarbonate nanocomposites: Rheology, conductivity and mechanical properties. Polymer. 2010;**51**:922-935. DOI: 10.1016/j.polymer.2009.12.041

[40] Gamba MM, Pouzada AS, Frontini PM. Impact properties and microhardness of double-gated glass-reinforced polypropylene injection moldings. Polymer Engineering & Science. 2009;**49**:1688-1695. DOI: 10.1002/pen.21393

[41] Gupta RK, Kennel E, Kim K-J, editors. Polymer Nanocomposites Handbook. Boca Raton, FL: CRC Press; 2009

[42] Costantino A, Pontes A, Pettarin V, Rosales C, Ramajo L, Viana J. Modification of PP with CNT as painting agent: Influence on performanceof injected pieces. In: Presented at the PMI2018 International Conference on Polymers and Mould Innovations, Guimarães, Portugal; 2018

[43] Jancar J, Dibenedetto AT. The mechanical properties of ternary composites of polypropylene with inorganic fillers and elastomer inclusions. Journal of Materials Science. 1994;**29**:4651-4658. DOI: 10.1007/BF00376292

[44] Jancar J, Dibenedetto AT. Failure mechanics in ternary composites of polypropylene with inorganic fillers and elastomer inclusions. Journal of Materials Science. 1995;**30**:2438-2445. DOI: 10.1007/BF01184598

[45] Ma CG, Mai YL, Rong MZ, Ruan WH, Zhang MQ. Phase structure and mechanical properties of ternary polypropylene/elastomer/nano-$CaCO_3$ composites. Composites Science and Technology. 2007;**67**:2997-3005. DOI: 10.1016/j.compscitech.2007.05.022

[46] Costantino MA, Rueda F, Pettarin V, Frontini PM, Pontes AJ, Viana JC. Characterization of PP/TPV/MMT ternary nanocomposites produced by injection molding. Macromolecular Symposia. 2017;**373**:1600153. DOI: 10.1002/masy.201600153

[47] Lim JW, Hassan A, Rahmat AR, Wahit MU. Morphology, thermal and mechanical behavior of polypropylene nanocomposites toughened with poly(ethylene-co-octene). Polymer International. 2006;**55**:204-215. DOI: 10.1002/pi.1942

[48] Dasari A, Zhang Q-X, Yu Z-Z, Mai Y-W. Toughening polypropylene and its nanocomposites with submicrometer voids. Macromolecules. 2010;**43**: 5734-5739. DOI: 10.1021/ma100633y

[49] Tiwari RR, Paul DR. Polypropylene-elastomer (TPO) nanocomposites: 1. Morphology. Polymer. 2011;**52**:4955-4969. DOI: 10.1016/j.polymer.2011.08.019

[50] Du H, Zhang Y, Liu H, Liu K, Jin M, Li X, et al. Influence of phase morphology and crystalline structure on the toughness of rubber-toughened isotatic polypropylene blends. Polymer. 2014;**55**:5001-5012. DOI: 10.1016/j.polymer.2014.08.012

[51] Kahl C, Feldmann M, Sälzer P, Heim H-P. Advanced short fiber composites with hybrid reinforcement and selective fiber-matrix-adhesion based on polypropylene—Characterization of mechanical properties and fiber orientation using high-resolution X-ray tomography. Composites Part A: Applied Science and Manufacturing. 2018;**111**:54-61. DOI: 10.1016/j.compositesa.2018.05.014

[52] Gamze Karsli N, Yesil S, Aytac A. Effect of hybrid carbon nanotube/short glass fiber reinforcement on the properties of polypropylene composites. Composites Part B: Engineering. 2014;**63**:154-160. DOI: 10.1016/j.compositesb.2014.04.006

Chapter 5

Tensile Properties in β-Modified Isotactic Polypropylene

Koh-hei Nitta and Tsutomu Takashima

Abstract

Spherulitic isotactic polypropylenes (iPPs) having a wide range of *β-phase* contents were prepared by adding *β*-nucleators, and the effects of the *β*-phase modification on the mechanical properties of the iPP were investigated. This chapter described the tensile properties of *β*-nucleated iPP, while key structural parameters, such as spherulite size and crystallinity, were controlled. The increase in the *β*-phase content led to broader yield peaks and an enhancement in the yield toughness but to a reduction in the yield strength. On the other hand, the initial elastic modulus was found to be independent of the *β*-contents. Furthermore, the deformation of the *β*-spherulites, which have a sheaflike structure, was anisotropic and depended on the stretching direction with respect to the sheaf axis. Consequently, the improved drawability and ductility of *β*-iPP compared to *α*-iPP are thus associated with the enhanced toughness resulting from the multiple deformation processes in the sheaflike spherulites.

Keywords: *β*-phase crystal, mechanical properties, tensile deformation, spherulite, crystalline morphology

1. Introduction

As well-known, isotactic polypropylene (iPP) is a polymorphic material with various crystal forms [1], such as monoclinic (α), hexagonal (β), triclinic (γ), and smectic, of which the *α*-phase is the most typical crystalline form. Commercial grades of iPP usually crystallize into the α-phase with sporadical occurrence of the *β*-phase under higher supercooling [2]. Crystallization under a temperature gradient [3] or flow-induced crystallization [4, 5] encourages the formation of the *β*-phase. To prepare of *β*-modified iPP samples, the introduction of selective *β*-nucleators is the most reliable method [6]. However, unless using specific *β*-nucleating agents, the *β*-phase cannot be obtained at high levels and is always accompanied by *α*-phase crystals. The *α*/*β* ratio is very sensitive to the crystallization temperature and the cooling rate because of the different nucleation rates of the two crystalline species. Varga et al. [7] found that pure *β*-phase can be achieved in the presence of some selective *β*-nucleators by the selection of appropriate thermal conditions for crystallization. Furthermore, the *β*-phase was found to be transformed to the *α*-phase by heat treatment [8]. This demonstrates that the monoclinic structure is thermodynamically stable, whereas the hexagonal *β*-phase is metastable and difficult to obtain under normal processing conditions.

Recently the number of practical studies has increased [9] because the impact strength and toughness of β-nucleated iPP exceed those of α-iPP. Although many studies have compared the mechanical properties of α-iPP and β-iPP, the morphological origin of the differences in the mechanical properties has not been clarified yet.

The mechanical properties of semicrystalline polymers such as iPP and polyethylene (PE) are governed by their morphological features which are specified by several structural variables such as the degree of crystallinity, spherulite size, crystalline thickness, and structural organization of the supermolecular structure [10]. These diversity and independencies of these structural variables make it difficult to provide a molecular or structural interpretation for the mechanical properties and deformation behavior of semicrystalline polymers [10]. Indeed, changing the thermal or processing conditions involves the concomitant modification of several structural parameters; thus, it is difficult to determine the structural origin of the change in mechanical properties as reported by Labour et al. [11]. Consequently, it is necessary to keep all the other structural parameters to be fixed to elucidate the effects of a given structural parameter on the mechanical properties. Very few studies have dealt with the mechanical properties of β-nucleated iPP with a wide range of β-phase contents, while all the other structural parameters, such as supermolecular organization and crystallinity, are controlled. The aim of this chapter is to elucidate the influence of the β-phase modification on the tensile properties of iPP. For this purpose, crystallization procedures, for the production of iPP sheets having a wide range of β-phase contents with fixed crystallinity and spherulite size, were developed. In addition, the effects of spherulitic morphology on tensile properties were investigated by comparing the differences in deformation mechanism of isolated α- and β-spherulites.

2. Structural characterization of β-modified isotactic polypropylene

2.1 Preparation method of β-modified polypropylene sheets

The starting material was a commercial iPP with a high tacticity (98%) obtained in powder form. The weight-averaged molecular weight M_w and polydispersity index M_w/M_n determined by gel permeation chromatography were 204×10^3 and 6.2, respectively. A β-nucleator prepared from an alcohol solution of pimelic acid and calcium stearate was used. The iPP powder was added to the solution, and it was dried in an oven at 373 K for 90 min.

The β-nucleator-added iPP powder was pressed and melted at 483 K. The samples were completely melted for 5 min between two aluminum plates prior to the application of 7.8 MPa pressure to produce specimens of approximately 0.2 mm thick. Adjustment of the degree of crystallinity in the volume fraction and the spherulite size to the same value for all the samples was made by changing the quenching temperature and the amount of the β-nucleators. The molten samples were allowed to equilibrate under pressure for 5 min prior to cooling. On removal from the press, the samples were plunged directly into a water bath maintained at an appropriate temperature 0, 30, 60, and 100°C and subsequently tempered for various periods at 100°C. Consequently, these procedures enabled us to achieve sample specimens having a wide range of β-phase contents with a constant crystallinity of about 65% and spherulite radius R of around 4 μm. The structural and morphological characteristics of the samples were summarized in **Table 1**. The end numeral of the sample code PP denotes the β-phase concentration in the crystalline

Sample	β -Cont. / %	ρ_c / kg m^{-3}	ρ / kg m^{-3}	χ_v / %	R / μm
PP0	0	936.0	908	65.9	3.6
PP16	16.4	933.5	906	65.3	3.7
PP43	42.7	929.6	904	66.1	4.3
PP65	65.4	926.2	901	65.1	5.1
PP83	83.0	923.5	900	66.1	4.9
PP98	97.8	921.3	898	65.3	4.4

Table 1.
Characteristics of iPP sheets.

fraction of the iPP. In this chapter, PP0 is denoted by α-iPP and PP98 is denoted by β-iPP.

The crystalline β-phase content (the volume fraction of the β-phase in the crystalline portion) was determined from WAXD patterns. The WAXD measurements were carried out at room temperature with a Rigaku RU-200 diffractometer with Ni-filtered Cu-Kα radiation from a generator operated at 40 kV and 100 mA. The β-phase fraction in the crystalline part of the specimens was assessed from the ratio of the area of the main (300) β-phase to the sum of the areas of the four main crystalline reflections: (110), (040), and (130) from the α-phase plus (300) from the β-phase using Turner-Jones method [12].

Here, we modified the analysis method proposed by Somani et al. [13] to obtain the volume fraction of β-phase in the crystalline fraction quantitatively. The reflection peaks in the WAXD profiles were deconvoluted. In the WAXD profile, (110) at 14.1°, (040) at 16.9°, and (130) at 18.5° are the principal reflections (in 2θ) of the α-phase crystals of iPP, whereas (300) at 16.1° is the principal reflection of β-phase crystals, and are considered as the markers for α-phase and β-phase crystals, respectively. The various reflection areas were computed after subtraction of the amorphous halo.

The volume fraction of β-phase crystals was calculated using the following equations:

$$\phi_\beta = \frac{S_{\beta(300)}\rho_{\beta c}^{-1}}{S_{\beta(300)}\rho_{\beta c}^{-1} + kS_\alpha \rho_{\alpha c}^{-1}} \tag{1}$$

and

$$S_\alpha = S_{\alpha(110)} + S_{\alpha(040)} + S_{\alpha(130)} \tag{2}$$

Here k is the calibration factor, $\rho_{\beta c}$ (=921 kgm^{-3}) is the density of the β-phase crystal [14–16], $\rho_{\alpha c}$ (=936 kgm^{-3}) is the density of α-phase crystal [17], S_β is the area of the (300) reflection peak, and S_α is the sum of the areas of (110), (040), and (130) peaks of α-phase crystals, respectively. The calibration constant k was estimated to be 1.11 from the difference in the sensitivity of WAXD reflections with respect to the thickness of the sheets for the α-phase and β-phase reflections.

Crystallinity can be precisely determined from density data. The densities of the specimens were determined by the flotation method. A binary medium prepared from various ratios of distilled water and ethanol was used. The volume crystallinity can be obtained using

$$\chi_V = \frac{\rho - \rho_a}{\rho_c - \rho_a} \tag{3}$$

where ρ is the overall density of the sample, ρ_a is the density of amorphous phase which was taken to be 854 kgm^{-3} [18], and ρ_c is the density of crystalline phase which was determined using

$$\rho_c = \phi_\beta \rho_{\beta c} + (1 - \phi_\beta)\rho_{\alpha c} \tag{4}$$

where ϕ_β estimated using Eq. (1) was employed.

2.2 Crystallization process

The morphological feature and the growth rate of the spherulites as a function of time were examined using a polarized optical microscope during the isothermal crystallization process. A polarized optical microscope (OLYMPUS, B201) fitted with an automated hot stage was used. The hot stage (METTLER TOLEDO, FP82HT) was held at a steady temperature to ± 0.2 K by a proportional controller. The film including β-nucleators was sandwiched between a microscope slide and a cover glass, heated to 483 K and kept at this temperature for 10 min to melt the crystallites completely. Then the samples were rapidly quenched to a given crystallization temperature T_c and allowed to crystallize isothermally. In **Figure 1**, the growth rates of α- and β-spherulites are plotted against the inverse of temperature. The growth rates increased with decreasing temperature over the whole experimental temperature ranges. **Figure 1** reveals that the difference between the two growth rates decreases with increasing temperature as shown by previous studies [6] and the growth rate of the β-spherulites exceeds that of the α-spherulite below 410 K which is slightly lower than 413–414 K determined by Shi et al. [19] and Varga [20]. This strongly suggests that the β-spherulites are relatively larger than the α-spherulites in iPP materials containing both phases prepared under usual conditions.

According to several kinetic theories [21–23], the growth rate G can be expressed by

$$G = G_0 \exp\left(-\Delta H/RT_c\right) \exp\left(-K_g/T\Delta Tf\right) \tag{5}$$

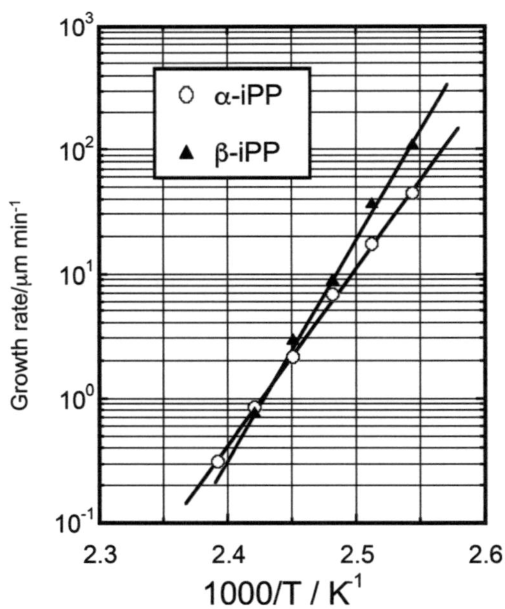

Figure 1.
Temperature dependence of growth rate of α- and β-spherulites.

where G_0 is a pre-exponential factor that is independent of temperature, ΔH, which is equal to 6.28 kJ/mol and corresponds to the activation energy of chain motion in the melt [22], $\Delta T = T_m^0 - T$ (T_m^0 is the equilibrium melting temperature), $f = 2T_c/(T_m^0 + T_c)$ is a correcting factor, and K_g is the nucleation constant in which the crystallization mechanism is divided into three regions, Regimes I, II and III, depending on the crystallization temperature and given by

$$K_{g(\mathrm{I})} = 2K_{g(\mathrm{II})} = K_{g(\mathrm{III})} = 4b_0\sigma\sigma_e T_m^0/\Delta h_f k_B \tag{6}$$

where Δh_f is the heat of fusion, b_0 is the thickness of the new layer, σ is the lateral surface free energy, σ_e is the fold surface free energy, and k_B is the Boltzmann constant. According to Hoffman et al. [21], the surface free energy can be estimated using

$$\sigma = \Delta h_f \frac{a_0 l_b}{l_u C_\infty} \tag{7}$$

where a_0 is the width of new layer, l_b (=0.154 nm) is the bond length, l_u (=0.1084 nm) is the projection length per atom, and C_∞ (=5.7) is the characteristic ratio [23]. The essential parameters for the kinetic study of β crystallization [19] are T_m^0 = 449 K, Δh_f = 177 MJm^{-3}, a_0 = 0.636 nm, and b_0 = 0.551 nm. As a result, the surface energy σ was estimated to be 1.4×10^{-4} J m^{-2}.

Using Eq. 6, $logG + \Delta H/2.303RT_c$ was plotted against $1/T\Delta Tf$ as shown in **Figure 2**. Two linear parts corresponding to Regime II and Regime III were obtained: the change in the slope occurs at 401 K, which is in the range (396–403 K) published in the literature [1], and the slope ratio is 1.74. It was estimated that σ_e = 2.68–3.08 $\times 10^{-2}$ J m^{-2} from the slopes using σ of 1.4×10^{-4} J m^{-2}. The work of chain folding q can be derived from the fold surface energy given by $q = 2\sigma_e a_0 b_0$. Consequently, the value of q for the β-phase was estimated to be 11–13 kJ/mol, which is about half the value (28 kJ/mol) for the α-phase given by Shi et al. [19].

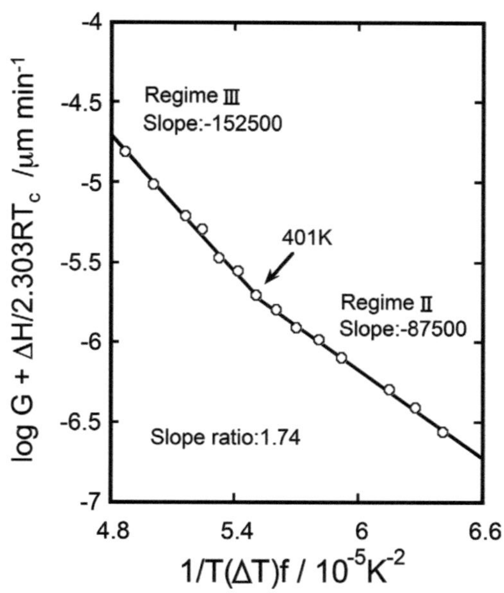

Figure 2.
Regime analysis of the growth rate of the β-spherulites.

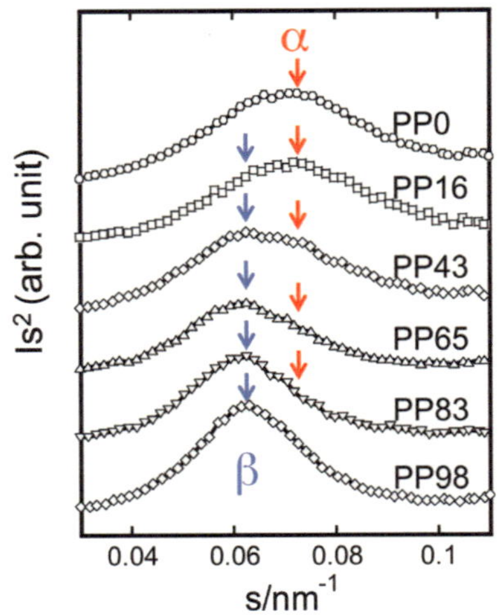

Figure 3.
Lorentz-corrected SAXS patterns of iPP samples having different β-contents with a fixed crystallinity.

2.3 Crystalline morphology

The small angle X-ray scattering (SAXS) measurement was performed with a point-focusing optics and a one-dimensional position-sensitive proportional counter (PSPC) with an effective length of 10 cm. The CuKα radiation supplied by a MAC Science M18X generator operating at 40 kV and 30 mA was used throughout. The distance between the sample and PSPC was about 40 cm. The geometry was further checked using a chicken tendon collagen, which gives a set of sharp diffraction spots corresponding to 65.3 nm.

From the volume fraction of the crystals χ_V, and SAXS long period L_p, the lamellar crystal thickness L_c and amorphous layer thickness L_a can be determined, assuming a two-phase model, from the following relationship:

$$L_c = \chi_V L_p, \qquad L_a = (1 - \chi_V) L_p \tag{8}$$

Figure 3 shows the Lorentz-corrected SAXS intensities plotted against magnitude of scattering vector s (= $2/\lambda \sin\theta$) where 2θ is the scattering angle and λ is the X-ray wavelength (= 0.1542 nm). The maximum point in the SAXS curves yields the average long period. The s value of α-PP (or PP0) was around 0.07 nm^{-1}, and the *s value* of β-iPP (or PP98) was around 0.0625 nm^{-1}, indicating that the long period L_p of β-iPP is greater than that of α-iPP. The iPP samples with both modifications have two SAXS peaks corresponding to the α-phase peak near 0.072 nm^{-1} and β-phase peak near 0.0625 nm^{-1}. This strongly suggests that the modified iPP samples with the α- and β-spherulites coexist but no co-crystallization of α-phase and β-phase crystals takes place. The specific long periods for α-phase and β-phase were about 14 and 16 nm.

3. Tensile deformation

The sample specimens were cut into a dumbbell shape having a gauge length of 10 mm. The tensile strain was calculated from the ratio of the increment of the length between the clamps to the initial gauge length. The tensile stress was determined by dividing the tensile load by the initial cross section. The stress–strain curves at room temperature were measured at a constant crosshead speed of 20 mm/min.

Figure 4a shows the overall stress–strain curves for all the samples with various β-phase contents at the same crystallinity. The ultimate tensile elongation markedly increases with increasing the β-phase content, and the β-iPP (PP98) has higher drawability than α-iPP (PP0). The β-iPP is elongated more gradually with ambiguous necking as compared to α-iPP, which is elongated with obvious necking.

As seen in **Figure 4b**, the initial elastic strain domain is surprisingly insensitive to the change in the composition of the crystalline phase at a fixed crystallinity. Thus, Young's modulus was constant and completely independent of the β-phase content (see **Figure 5**). This phenomenon is responsible for the strain concentration in the amorphous region [24] because the amorphous phase in iPP is rubberlike at room temperature and the mechanical modulus of the amorphous phase is considerably lower than those of α- and β-phase crystals. Consequently, before yielding, the deformation of the semicrystalline polymers is dominated by the deformation of the amorphous phase, indicating that the initial elastic region depends mainly on the crystallinity.

The elasticity limits where the actual stress-strain curves for the β-modified iPP samples are deviated from the linear elastic behavior were around 0.1 strain as shown in **Figure 4b**. The deviation may be due to the onset of microscopic plastic

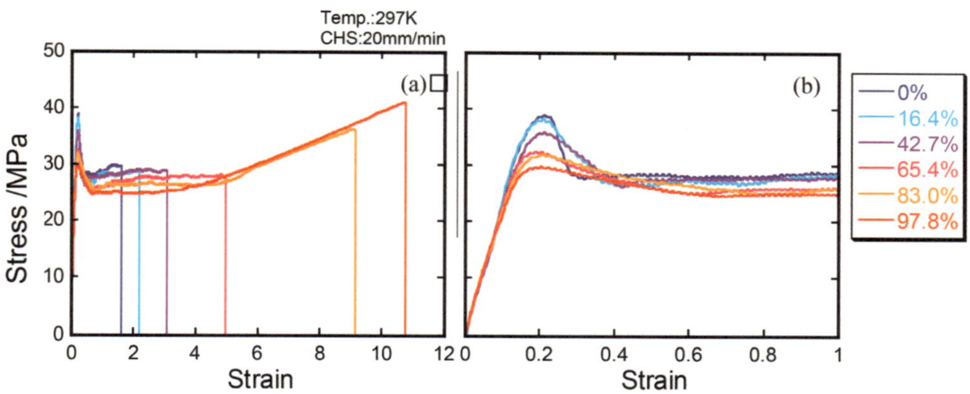

Figure 4.
Stress-strain curves of iPP samples having different β-contents with a fixed crystallinity. (a) Overall curves and (b) their magnification in the initial strains.

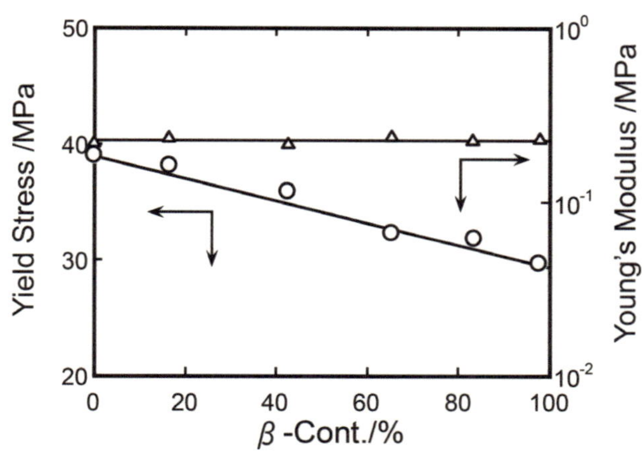

Figure 5.
Yield stress and Young's modulus plotted against the β-contents for the β-nucleated iPP.

deformation resulting from the lower packing density of the β-phase. The main differences in the stress-strain curves exist in the yield region. In this region, the macroscopic structural transformation from a spherulitic structure to microfibrils takes place.

The yield process was found to become broader as the β-phase content increases, and the yield stress linearly decreases with increasing β-phase content (see - **Figure 5**). This is relevant to the early and more gradual activation of plastic processes in the β-phase as compared to α-phase because of the higher molecular mobility in the β-phase at the same temperature. This demonstrates that the plastic behavior is much more sensitive to the nature of the crystalline phase. In addition, the yield peak in the stress–strain curves broadens, and the neck region is more ambiguous as the β-phase increases. As mentioned before, the β-phase crystals have a lower cohesive force than the α-phase crystals, which is also reflected by their lower melting temperature and lower density. The lower cohesive force leads to easier slipping of the lamellar chains, resulting in a lower yield stress. Furthermore, as shown in **Figure 6**, the yield energy, which is defined as the energy dissipated for yielding to take place, linearly increases with increasing β-phase content, and all data almost fall on the solid line, which can be calculated according to the simple mixture law as follows:

$$U_Y = \phi_\beta U_{\beta Y} + (1 - \phi_\beta)\ U_{\alpha Y} \tag{9}$$

Here U_Y is the yield energy (resilience), which was estimated from the area under the stress-strain plot from the origin to the stress drop, and $U_{\beta Y}$ and $U_{\alpha Y}$ are the yield energies of β-iPP and α-iPP, respectively.

To obtain better insights into the plastic behavior of the crystalline component, the WAXD experiments were carried out at room temperature during tensile tests. The direction of the incident beam was perpendicular to the plate surface of the specimens. **Figure** 7 shows the WAXD patterns of α-iPP (PP0) and β-iPP (PP98). The WAXD patterns of the undrawn specimens of PP0 and PP98 are shown in **Figures 7a** and **d**, respectively. The patterns of iPP0 stretched at a strain of 0.4 and PP98 stretched at a strain of 0.8, in which both stretched samples are in the post-yielding region, are shown in **Figures 7b** and **e**, respectively, and those of both specimens at the final failure point are shown in **Figures 7c** and **f**, respectively. The Debye rings of the (300) and (030) reflections of the β-phase crystals remain in the

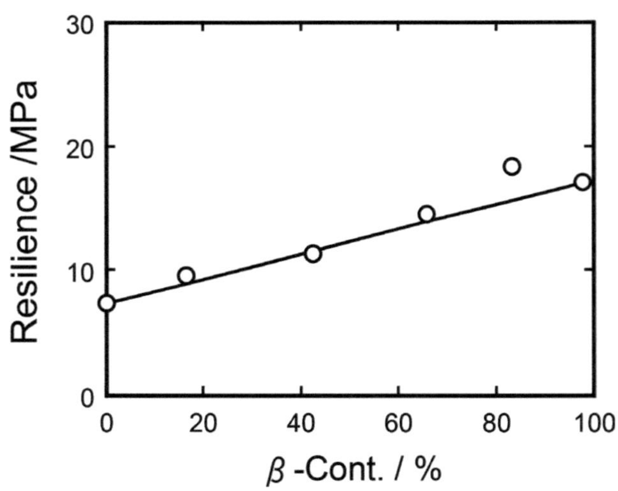

Figure 6.
Yield energy (resilience) plotted against the β-contents for the β-nucleated iPP.

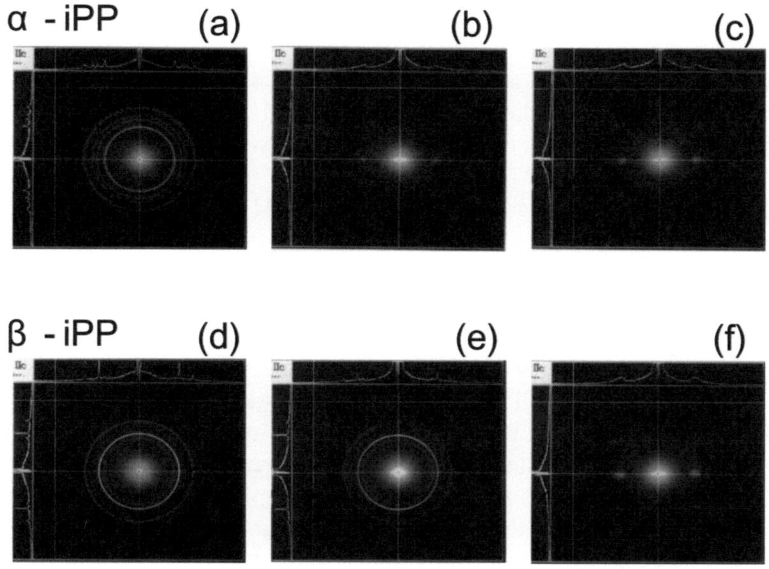

Figure 7.
Small angle X-ray diffraction patterns of α- and β-iPP samples: (a) original α-iPP, (b) stretched α-iPP at a strain of 0.4 (neck region), (c) stretched α-iPP at the failure point, (d) original β-iPP, (e) stretched β-iPP at a strain of 0.8 (neck region), and (f) stretched β-iPP at the failure point.

PP98 sample deformed up to the post-yielding region (see **Figure 7e**), and they are scarcely oriented even in the initial necking region. In contrast to PP98, PP0 exhibits the reflections of the α-phase concentrated in the perpendicular direction to the elongation even in the post-yielding and necking regions. This pattern is almost the same as the typical profile of the iPP specimens at the final failure point, indicating that α-iPP attains the final c-axis orientation of crystals after yielding. In addition, it should be noted here that there is no clear difference in the orientation pattern at the final failure point between PP98 and PP0. The strain induced $\beta \to \alpha$ transition on tensile drawing has been reported by several authors [25–29]. In the present work, there is no evidence for the occurrence of a $\beta \to \alpha$ transition as seen in **Figure 7f**, but the final orientation morphology of PP98 appears to be the same as that of PP0. The reflections showing the attainable final orientation exist between 14 and 16°, suggesting the assignment of smectic form as demonstrated by Turner-Jones et al. [12] and Shi et al. [8].

4. Deformation of isolated spherulites

According to our previous studies [30, 31] concerning the yield behavior of typical spherulitic polymers such as PE and α-iPP, several lamellae tend to cluster into bundles with tie molecules, where these are separated from one another by the amorphous regions and the lamellar clusters constituting of spherulites act as deformation units. The lamellar clusters are bridged by the inter-cluster or intercrystalline links, as proposed by Keith-Padden et al. [32], thus acting as stress transmitters. The stacked lamellae or lamellar clusters are fragmented into cluster units or blocks at the yield point, resulting in a stress drop. Beyond the yield point, the plastic deformation involves the rotation of the cluster units and the sliding of stacked lamellae inside each cluster units, and the fragmented cluster units are rearranged into microfibrils in the necking region [33]. The continuous structural transformation corresponds to the neck propagation. In the case of β-spherulitic iPP showing a broad yield process, the lamellar clusters disintegrate accompanied by sliding crystalline stems and chain slip inside the crystalline lamellae. Consequently, the fragmentation of lamellae and/or lamellar clusters hardly occurs. These differences in yielding mechanism between α- and β-spherulitic iPPs are due to the differences not only in the cohesive force between crystalline chains but also in the spherulite morphology.

Figure 8 shows an atomic force microscopy (AFM) micrograph of the morphology of the β-form iPP spherulites prepared in this work. The morphology is significantly different from the typical spherulite morphology of α-phase iPP. This spherulite seems to be type III according to Norton and Keller's classification [34] because there is no lamellar twisting within the spherulites. The embryo β-spherulites consist of parallel stacked lamellae. This type of spherulite is referred to as a sheaflike structure. Shi et al. [19] reported that the β-spherulites develop initially as rodlike structures and then by branching of the lamellae, finally evolving into sheaflike structures. In this case, the spherulite is formed from one crystal via a unidirectional growth mechanism. The spherical shape is attained through continuous branching and fanning via the intermediate stage of sheaves. However, the α-spherulites consist of an aggregate of chain-folded lamellae growing from a central point (nucleus). This is referred to as an acicular structure. Both structural models of α- and β-spherulites are shown in **Figure 9**.

It is likely that the spherulite morphology plays a central role in controlling the plastic deformation and tensile behavior of both PE and iPP materials. The mechanical responses to tensile yielding and the deformation process are considered to be fundamentally different between α- and β-spherulites. Tensile deformation of α-iPP materials is accompanied by necking process, in which the initial isotropic

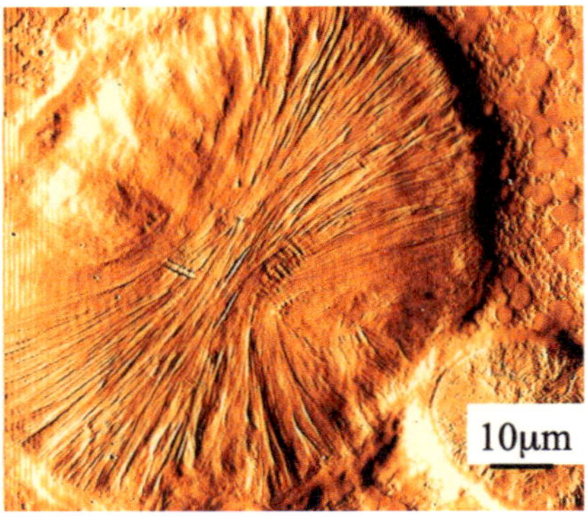

Figure 8.
AFM pictures of a β- spherulite.

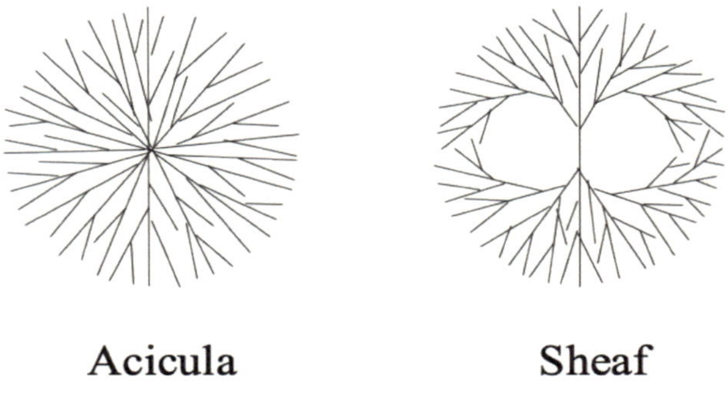

Figure 9.
Illustrations of lamellar arrangement of α- and β-spherulites.

spherulite structure is converted into a highly oriented one. On the other hand, the β-iPP exhibits broader yield peaks without obvious necking formation and has a lower yield strength than that of α-iPP. Furthermore, the β-iPP specimen began to be whitened with further extension, whereas the α-iPP exhibited partial stress whitening beyond yielding. This stress whitening seems to be caused by the formation of numerous voids after yielding [35].

Previously, we reported a method for preparing a thin film with huge isolated spherulites embedded in a soft (smectic) matrix [36]. The deformation mechanism of spherulites can be examined from the direct observation of the stretched film on the polarized optical microscope. Thus, we mounted the manual stretcher on the optical microscope to observe the deformation process of the isolated spherulites.

Figure 10a shows the optical microscopic pictures of an isolated α-spherulite. A few arc-shaped cracks rapidly appeared in the polar zone in the initial stages of

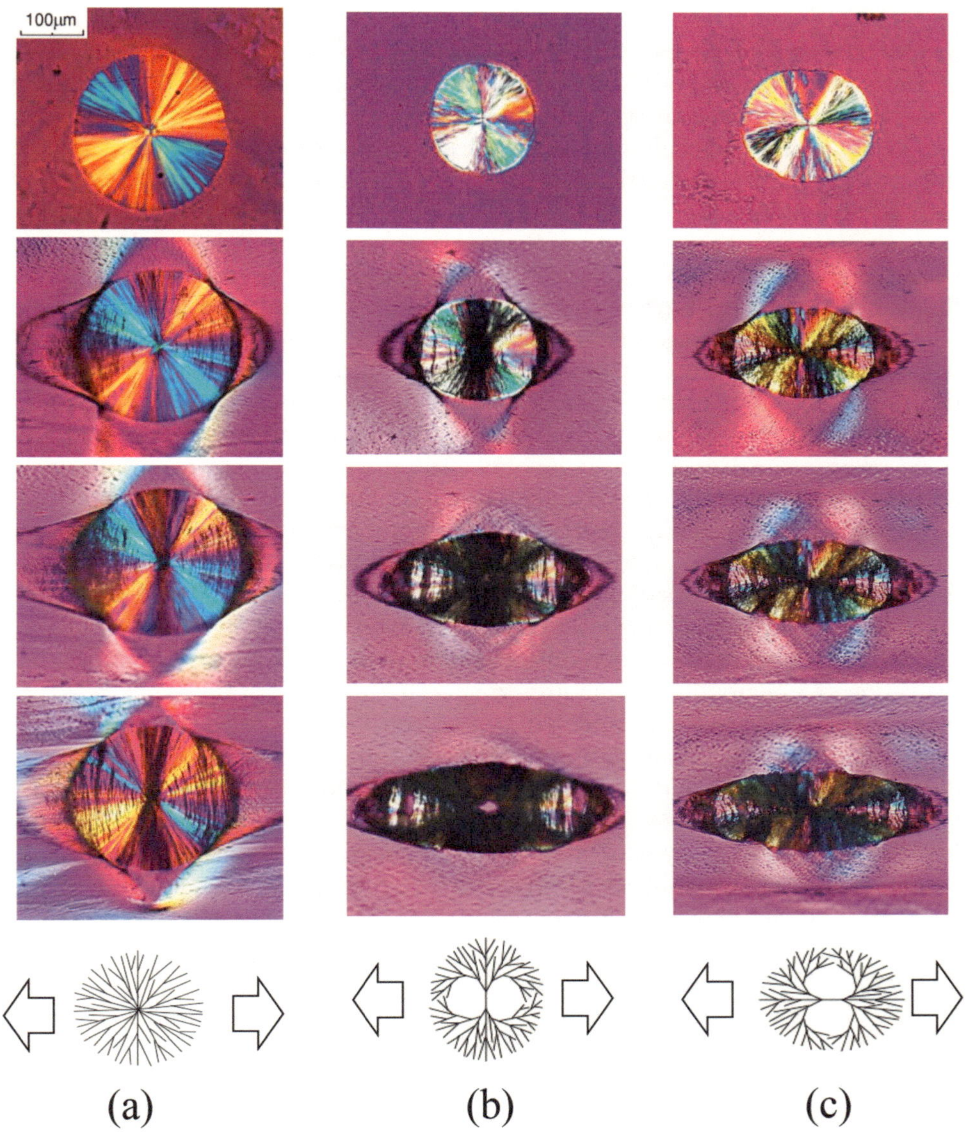

Figure 10.
Polarized optical microscopic pictures of (a) uniaxial stretching of an isolated α-acicular spherulite, (b) uniaxial stretching perpendicular to the sheaf axis of an isolated β-sheaf spherulite, and (c) uniaxial stretching parallel to the sheaf axis of an isolated β-sheaf spherulite.

stretching. With increasing strain, the arc-shaped cracks developed in the polar zone and proceed from the outer to the inner portions of the spherulite. Subsequently, radial craze-like fractures began to form in the equatorial region perpendicular to the stretching axis, and then the radial crazing progressed along with the spherulite radius, resulting in the evolution of large dark bands in the equatorial region. The evolution of the dark bands is related to the yield process as demonstrated previously by Nitta et al. [36]. It should be noted here that the deformation mechanism of α-spherulites is isotropic because crystalline lamellae within α-spherulites radiate from a common center and the crystalline lamellae aggregate with spherical symmetry. Unlike the α-spherulite, β-spherulites are sheaflike type of spherulites with a spherical asymmetry (see **Figure 9**). As shown in **Figure 10b** and **c**, the deformation behavior of the β-spherulite depends largely on the stretching direction with respect to the sheaf axis. In the case of the β-spherulites, when drawn perpendicular to the sheaf direction, the radial crazing preferentially appeared in the equatorial zone along the sheaf axis, and then the dark crazing zone developed further with increasing strain. Finally, a hole appeared in the center of the deformed spherulite, indicating that the deformation is concentrated perpendicular to the stacked lamellae located in the center of the spherulite. On the other hand, when drawing along the sheaf axis, the spherulite was deformed into an ellipsoid accompanied by the formation of crazed cracks, and there is no clear strain concentration.

It was found that the strength of the β-spherulite is anisotropic and depends on the direction of the embryo or parallel stacked lamellae in the center of the spherulite. When the β-spherulites were subjected to stress perpendicular to the sheaf axis (see **Figure 10b**), obvious deformation bands generated preferentially near the equatorial zone within the uniaxially deformed spherulites. According to previous theoretical [37, 38] and experimental results [39], the equatorial region, particularly the center of the spherulites, is subjected to higher strains and stresses as compared to the polar region. Consequently, interlamellar separation is likely to occur near the equatorial plane of the stacked sheaflike lamellae because the sheaf direction is perpendicular to the loading direction. As the strain increased, separation of the sheaf-lamellae continued, and more deformation bands and crazes generated preferentially near the equatorial zone of the deformed spherulites. In the final stage, holes or local disintegration appeared near the center of the deformed spherulites. This lamellar separation was accompanied by massive voiding at the onset of the formation of a microporous structure, which is preferential for the applications of β-phase iPP [29, 40]. On the other hand, when the β-spherulite was stretched in the growth direction of the embryo sheaf (see **Figure 10c**), there was no obvious deformation bands around the equatorial zone. Thus, intralamellar deformation is likely to take place for the sheaf-lamellae under uniaxial tension because the sheaf-lamellae are parallel to the loading direction. Considering that the intralamellar stretching of the sheaf-lamellae involves the unfolding of chains, leading to local necking or sliding, the intralamellar stretching of sheaf-lamellae strongly resists deformation compared to the interlamellar separation; thus, no localized deformation bands appeared near the equatorial zone.

As well-known, the $\beta \rightarrow \alpha$ transformation occurs on heat treatment. The film having isolated β-spherulites was heated up to 433 K at a rate of 2 K/min and then quenched in an ice-water bath. This treatment allows the recrystallization into α-modification within the isolated β-spherulites. The arrangement of the crystalline lamellae in the α-spherulites prepared by the $\beta \rightarrow \alpha$ transformation is a sheaflike structure, which is different from the usual α-spherulites showing an acicular structure. Thus, the sheaflike spherulite prepared by the $\beta \rightarrow \alpha$ transformation process is a new type of α-spherulite. As shown in **Figure 11**, the sheaflike

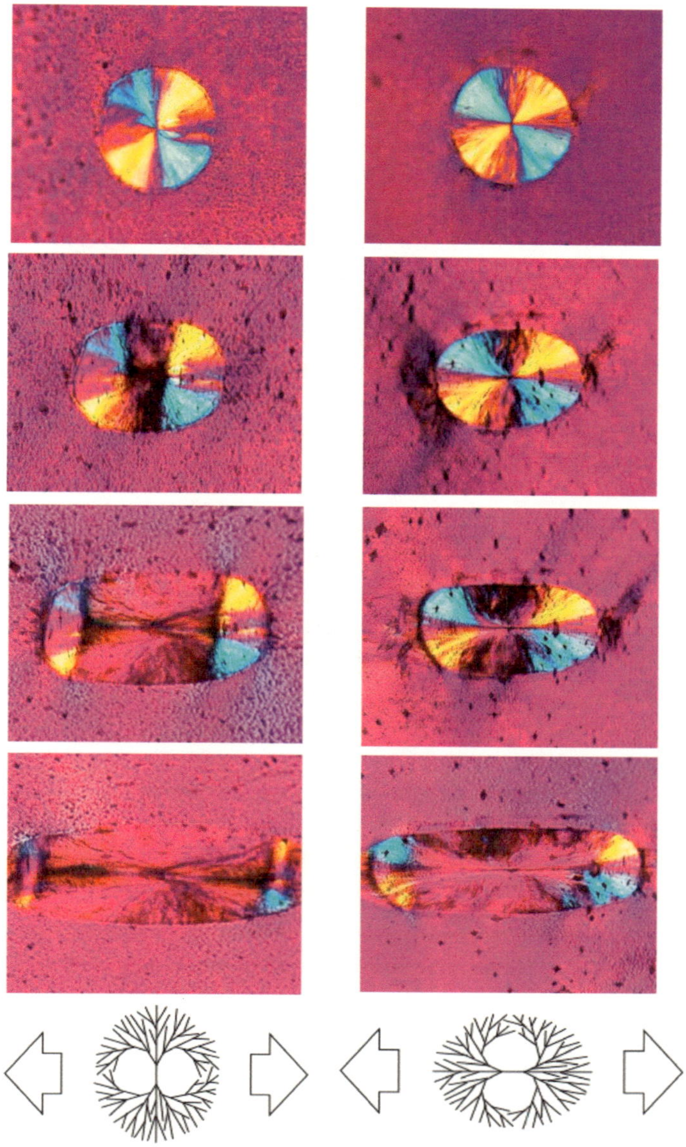

Figure 11.
Polarized optical microscopic pictures of (a) uniaxial stretching perpendicular to sheaf axis of an isolated α-sheaf spherulite and (b) uniaxial stretching parallel to sheaf axis of an isolated α-sheaf spherulite.

α-spherulite is optically negative, indicating that there are no traces of a cross-hatched structure within the β-spherulites.

The deformation behavior of the sheaflike α-spherulite was also anisotropic and significantly different from that of the acicular type of α-spherulite as shown in **Figure 10**. When the axis of the sheaf was transverse to the loading direction, the deformation bands appeared obviously along the sheaf axis, and then the further deformation extended the highly oriented and deformed zone in the equatorial region of the deformed spherulite. The uniaxially deformed spherulite is clearly divided into two parts: one being nearly undeformed and another being considerably deformed. The nearly undeformed sections are jointed by the transition zone that propagates in the stretching direction. The interlamellar separation of stacked sheaf-lamellae in the equatorial zone was initiated in the first stage of deformation, and then the separation of the sheaf-lamellae continues, and more deformation bands proceeded as the strain was further increased. On the other hand, when the sheaf axis was in the draw direction (see **Figure 11b**), the spherulite was initially deformed to an ellipse of similar shape to that expected for affine deformation. This is because its equatorial region is tougher because the lamellae parallel to the stretching direction more strongly resist deformation than the lamellae perpendicular to the loading direction as mentioned in the discussion of β-spherulite deformation. Subsequent deformation caused micro-necking in such a way that the traces of the sheaf structure remain in the center portion of the spherulite. The sheaf-lamellae located perpendicular to the loading direction are brittle, whereas the sheaf-lamellae located parallel to the loading direction are tougher or ductile. This anisotropic deformation behavior is quite different from the isotropic deformation of acicular α-spherulites, but it is similar to those of sheaf β-spherulites as well as isolated PE spherulites, as shown by Lee et al. [41]. This is plausible because PE spherulite is sheaflike.

It has been long recognized that the deformation of crystalline polymers must be considered in terms of various structural parameters such as crystallinity, lamellar thickness or long period, and spherulite size. However, the present results imply that deformation behavior and mechanical response of bulk iPP materials are affected not only by these structural factors but also by the morphological texture within spherulites.

5. Effects of spherulite morphology on tensile properties

To investigate the effects of the lamellar organization within the spherulites on the tensile properties of bulk iPP sheets, α-modified iPP samples were prepared by

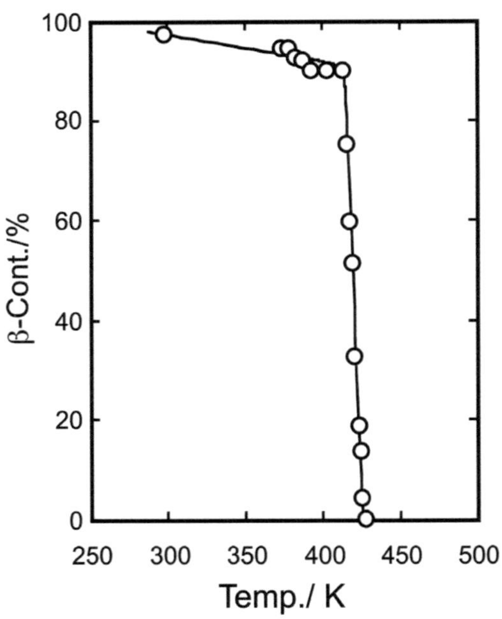

Figure 12.
Dependence of β-contents of β-nucleated iPP on the heat-treatment temperature.

the heat treatment of β-iPP (PP98) sheets. The PP98 sheets were heated at a 2 K/min and kept for 300 min at a fixed temperature. The β-phase contents are plotted against the fixed temperature in **Figure 12**. The $\beta \rightarrow \alpha$ transformation occurs at around 413 K, and the β-iPP was completely transformed into the α-phase above 427 K. It should be noted here that the thus-prepared iPP sheets contain sheaf type of spherulites. Consequently, we obtained three types of iPP sheets having a fixed crystallinity of around 73%, for example, the α-iPP sheets showing acicular spherulites, the α-iPP sheets showing sheaflike spherulites, and the β-iPP sheets showing sheaflike spherulites. Here, we have referred to these samples as α-acicular, α-sheaf, and β-sheaf.

Figure 13 shows the stress-strain curves measured at various temperatures for α-acicular, α-sheaf, and β-sheaf sheets. At all temperatures, the stress–strain curves in the initial elastic strain domain were almost the same for these three samples. This is plausible because the crystallinities of these samples are almost equal. This also indicates that Young's modulus is dominated by the bulk crystallinity and is almost independent of the lamellar morphology of the spherulites and of the crystal modification. In addition, the α-acicular iPP sample is in more brittle manner than the α-sheaf and the β-sheaf iPP samples and broke around the yield peak except at 380 K. This indicates that the plastic deformation is much more sensitive to the change of the spherulite texture than to crystalline modification. This corroborates the previous results that the deformation behavior of isolated β-sheaf and α-sheaf spherulites is similar and significantly different from that of the α-acicular spherulites. Moreover, β-spherulites show a greater resistance to break when the strain direction is almost parallel to the sheaf axis. Interestingly, the yield strengths in α-acicular and α-sheaf iPPs are almost the same, although

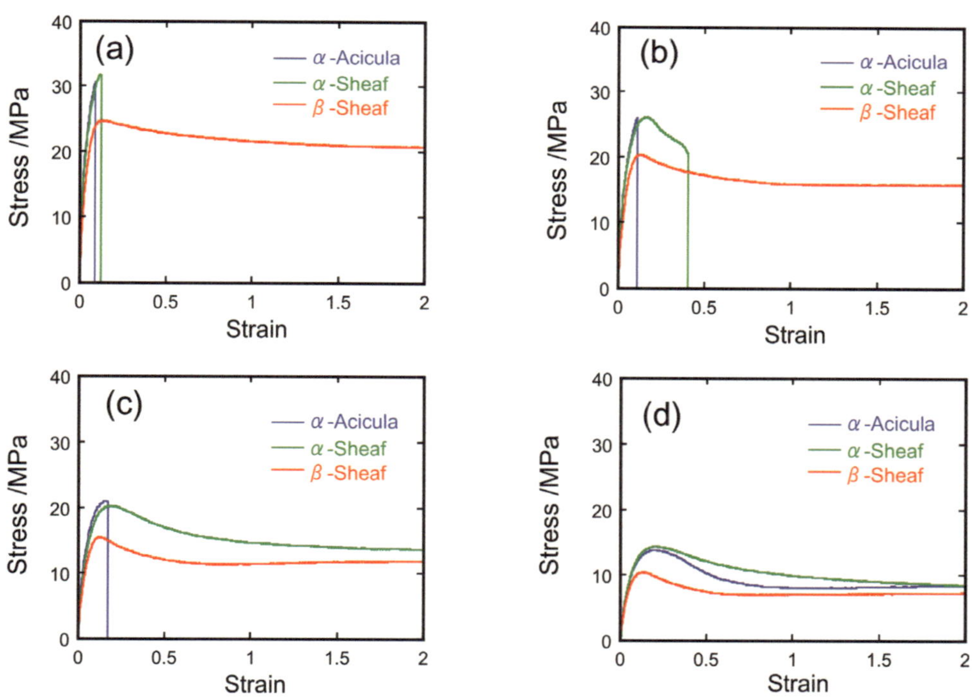

Figure 13.
Comparison of stress-strain curves of spherulitic iPP sheets with a fixed crystallinity: α-acicular spherulites (blue), α-sheaf spherulites (green), and β-sheaf spherulites (red).

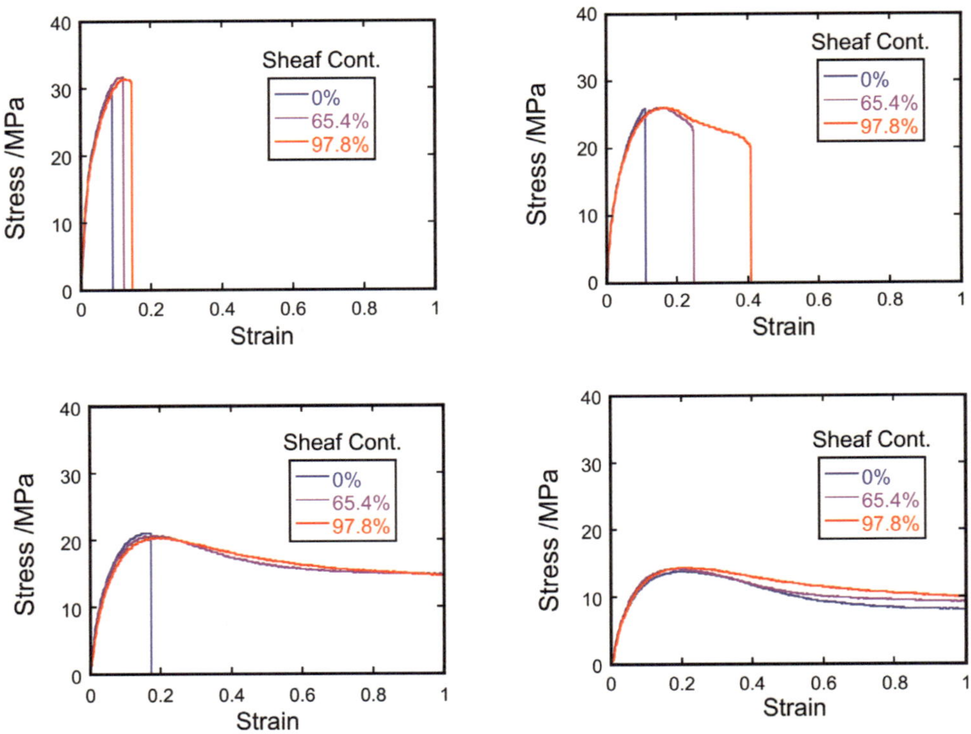

Figure 14.
Comparison of stress-strain curves of α-spherulitic iPP samples with different contents of sheaflike spherulites.

larger than that of the β-sheaf iPP. This indicates that the yield strength is much more sensitive to crystal phase modification than to the lamellar arrangement of the spherulites. The reduced yield stress of β-sheaf iPP compared to those of the α-acicular and α-sheaf iPPs is associated with the greater chain mobility in the β-phase crystals. The lower packing density of the β-phase is accompanied by the reduced stem interactions in the β-crystalline lamellae as compared to the α-lamellae, leading to the lower yield stress of the β-phase compared to that of the α-phase. An additional factor reducing the lamellar strength is the crystallographic symmetry of the hexagonal β-phase, which provides three equivalent glide planes.

To confirm these conclusions, we compared the stress-strain behaviors measured from 320 to 380 K for α-spherulitic iPP sheets with different amounts of sheaflike spherulites which were prepared by tempering the iPP samples with the different amount of β-spherulites. Note here that these iPP sheets have a fixed crystallinity of about 74%. As a result, α-spherulitic iPP sheets having various sheaflike spherulite contents with constant crystallinity were prepared. **Figure 14** also shows that the ductility enhances as the content of sheaf spherulites increased, whereas the yield strengths of all sheets are the same and almost insensitive to the lamellar arrangement of the spherulites.

6. Summary

Based on our investigation of the tensile properties of α-iPP and β-modified iPP, in which all the other structural parameters, such as overall crystallinity and spherulite size, were controlled, the following conclusions can be drawn:

(1) The stress-strain curves in the initial elastic strain region are dominated by the effects of crystallinity but are almost insensitive to changes in the crystal phases, as well as to the lamellar arrangements of spherulites.

(2) The yield strength is more sensitive to the crystal modification than the lamellar arrangement of spherulites, and α-iPP shows a higher yield stress than β-modified iPP.

(3) The plastic deformation process of the spherulites is sensitive to the lamellar arrangement of spherulites. The sheaflike spherulites are more ductile than the acicular spherulites and exhibit anisotropy in their plastic properties. The improved drawability and ductility of β-iPP compared with α-iPP is thus associated with the enhanced toughness resulting from multiple deformation processes in the sheaflike spherulites.

Author details

Koh-hei Nitta* and Tsutomu Takashima
Institute of Science and Engineering, Kanazawa University, Kanazawa, Japan

*Address all correspondence to: nitta@se.kanazawa-u.ac.jp

References

[1] Karger-Kocsis J. Polypropylene Structure, Blends and Composites. Vol. 1. London: Chapman and Hall; 1995

[2] Padden FJ Jr, Keith HD. Journal of Applied Physics. 1959;**30**(10): 1479-1484

[3] Lovinger AJ, Chua JO, Gryte CC. Journal of Polymer Science Part B: Polymer Physics. 1977;**15**(4):641-656

[4] Varga J. Karger-Kocsis. Journal of Polymer Science Part B: Polymer Physics. 1996;**34**:657-670

[5] Varga J, Ehrenstein GW. Polymer. 1996;**37**(26):5959-5963

[6] Varga J. Crystallization, Melting and Supermolecular Structure of Isotactic. In: Karger-Kocsis J, editors. Polypropylene: Structure, Blends, and Composites. Vol. 1. London: Chapman and Hall; 1995. pp. 56-115

[7] Varga J, Mudra I, Ehrenstein GW. Journal of Applied Polymer Science. 1999;**74**(10):2357-2368

[8] Shi G, Zhang X, Cao Y, Hong J. Makromolekulare Chemie. 1993;**194**(1): 269-277

[9] Varga J, Breining A, Ehrenstein GW, Bodor G. International Polymer Processing. 1999;**14**(4):358-364

[10] Kennedy MA, Peacock AJ, Mandelkern L. Macromolecules. 1994; **27**(19):5297-5310

[11] Labour T, Gauthier C, Seguela R, Vigier G, Bomal Y, Orange G. Polymer. 2001;**42**(16):7127-7135

[12] Turner Jones A, Aizlewood JM, Beckett DR. Makromolekulare Chemie. 1964;**75**:134-158

[13] Somani RH, Hsiao BS, Nogales A, Fruitwala H, Srinivas S, Tsou AH. Macromolecules. 2001;**34**(17): 5902-5909

[14] Keith HD, Padden FJ, Walter NM, Wyckoff HW. Journal of Applied Physics;**1959**(10):30, 1485-1488

[15] Samuels RJ, Yee RY. Journal of Polymer Science Part A2. 1972;**10**: 385-432

[16] Addink EJ, Beintema J. Polymer. 1961;**2**:185-193

[17] Natta G, Corradini P. Nuovo Cimento. Suppl. 1960;**15**(S1):40-51

[18] Newman S. Journal of Polymer Science, Polymer Physics Edition. 1960; **47**:111

[19] Shi G, Zhang X, Qiu Z. Makromolekulare Chemie. 1992;**193**(3): 583-591

[20] Varga J. Journal of Materials Science. 1992;**27**(10):2557-2579

[21] Hoffmann JD, Miller RL. Macromolecules. 1988;**21**:3038-3051

[22] Hoffmann JD. Polymer. Vol. 231982. pp. 656-670

[23] Hoffmann JD, Miller RL. Polymer. 1997;**38**:3151-3212

[24] Nitta KH, Yamaguchi N. Polymer Journal. 2006;**38**(2):122-131

[25] Asano T, Fujiwara Y. Polymer. 1978; **19**(1):99-108

[26] Fujiyama M. International Polymer Processing. 1999;**14**:3-9

[27] Karger-Kocsis J, Varga J. Journal of Applied Polymer Science. 1996;**62**(2): 291-300

[28] Li JX, Cheung WL. Polymer. 1998; **39**(26):6935-6940

[29] Chen HB, Karger-Kocsis J, Wu JS, Varga J. Polymer. 2002;**43**(24): 6505-6514

[30] Nitta KH, Takayanagi M. Journal of Macromolecular Science, Part B: Physics. 2003;**42**:107-126

[31] Takayanagi M, Nitta K, Kojima O. Journal of Macromolecular Science, Part B: Physics. 2003;**42**:1049-1059

[32] Keith HD, Padden FJ, Vadimsky RG. Journal of Polymer Science Part A-2. 1966;**4**:267-281

[33] Kuriyagawa M, Nitta KH. Polymer. 2011;**52d**:3469-3477

[34] Norton DR, Keller A. Polymer. 1985; **26**(5):704-716

[35] Varga J. Journal of Macromolecular Science, Part B. Physics. 2002;41 (4–6), 1121-1171

[36] Nitta KH, Takayanagi M. Journal of Materials Science. 2003;**38**(24): 4889-4894

[37] Nitta KH. Computational and Theoretical Polymer Science;**1999**(1):9, 19-26

[38] Wang TT. Journal of Polymer Science Part B: Polymer Physics. 1974; **12**(1):145-158

[39] Rodriguez-Cabello JC, Alonso M, Merino JC, Pastor JM. Journal of Applied Polymer Science. 1996;**60**:1709-1717

[40] Coulon G, Castelein G, G'Sell C. Polymer. 1996;**37**:2309

[41] Lee SY, Bassett DC, Olley RH. Journal of Materials Science. 2000; **35**(20):5101-5110

Chapter 6

Effect of Processing and Orientation on Structural and Mechanical Properties of Polypropylene Products

Luca Fambri and Luca Lutterotti

Abstract

Polypropylene (PP) represents one of the most worldwide used plastics with a large variety of products and applications. As usual for semicrystalline polymers, the properties of PP products strictly depend on the processing (fiber spinning, film extrusion, injection, etc.), where orientation and crystallization phenomena are involved. The object of this communication is the mechanical and structural characterization of oriented products from iPP homopolymers, i.e., injection molded dumbbell specimens (IM), lab-scale single fibers and commercial bulk continuous filament (BCF), woven non-woven fabrics (WNW) by using differential scanning calorimetry (DSC), dynamical mechanical thermal analysis (DMTA), tensile measurements, and X-ray diffraction (XRD) analysis. In particular, a recent methodology to analyze diffraction images of oriented polymers to obtain crystal structure, texture, and microstructural information is presented. The higher the orientation, the higher the mechanical properties and the sharper the texture, as revealed by a quantitative texture analysis that has been also developed and successfully applied to oriented PP nanocomposites.

Keywords: fibers, woven non-woven fabrics, tensile properties, DMTA, DSC, XRD analysis

1. Introduction

On the market volume basis of standard plastics, polypropylene occupies the second position after various polyethylenes (HDPE, LDPE, and LLDPE) and is expected to rise by 3% per year by 2024 according to the recent report by Ceresana [1]. Following PlasticsEurope about 9.9 m of PP were demanded in 2017 for food packaging, sweet and snack wrappers, hinged caps, microwave containers, pipes, automotive parts, bank notes, etc. in Europe [2]. The success of polypropylene derives from the proper balanced physical and chemical properties with a combination of many factors, such as low density, excellent thermal stability, good chemical resistance, high crystallinity, and high stiffness or hardness with a wide design flexibility and simplicity of recycling making PP an attractive construction material. Owing to plausible melt rheology and thermal properties, PP-based materials are compatible with many processing technologies and can be processed by injection molding, film blowing and casting, extrusion of woven non-woven fabrics,

IntechOpen

fiber spinning, and calendering [3, 4]. Also polypropylene composites for injection molding represent a sector of production for applications where higher stiffness and hardness are required. These higher mechanical properties are often obtained by addition and dispersion of particulate fillers and/or glass fibers [5].

Depending on the process and applications, various grades of polymer are properly selected and defined by melt flow index (MF) that represents the output of polymer expressed in g/10 mins during a vertical extrusion and measured in standard conditions, usually at 230°C and load of 2.16 kg with a diameter die of 2.096 mm [6]. @@@Melt flow grades in the range 0.3–2.0 are chosen for pipes, sheets, and blow molding, whereas higher grades between 2 and 8 are selected for film and fiber production. Higher fluidity polymers with MF above 8 are typically used for injection molding and extrusion coating [7]. Common MF for woven non-woven (WNW) fabrics is usually in the range 18–25.

During various processing conditions, polymer chains are oriented, and the final properties directly depend on the combined interaction of polymer crystallized and amorphous phases during shear flow and/or elongational flow. The latter is typically dominating in film and fiber production. In particular, it is possible to achieve a high extension of macromolecular chains due to the relatively easy spinning, with a very high level of alignment [8, 9]. According to Kunugi, experimental maximum modulus and strength of about 40 and 1.5 GPa, respectively, could be obtained for highly extended helix crystallizing polymer [10]. In the case of injection molded polypropylene, an extensive work has been published with specific details on the effect of macromolecular orientation on polymer structure, after process, characterized by XRD, DSC, and DMTA in dependence on various factors, such as skin layer, filler effect, copolymers, flow direction, processing parameters, etc. [11].

In this chapter both injection molded samples and fiber-based products are compared as function of polymer processing. Mechanical, thermal, and structural analyses are presented. Particular attention will be spent on interpretation of X-ray analysis on fiber oriented polymers, comparing results before and after mechanical and creep tests.

2. Experimental

2.1 Products and polymers

Various products of PPs with different MF in the range 3.6 and 18 g/10 mins were selected and analyzed for the comparative evaluation of the effect of process on properties and orientation. The higher the molecular weight, the lower the melt flow. **Figure 1** shows the empirical relation between molecular weight and melt flow according to literature data [12] and Eq. (1):

$$\log MW = 2.47\text{–}0.234 \log MF \quad (1)$$

where MW is the molecular weight expressed in kDalton and MF is the melt flow measured in standard conditions (230°C and 2.16 kg). The MF value of polypropylene of various products is also reported. See details in **Table 1**.

Dumbbell specimens ASTM type (3.2 thickness, 12.7 width) were obtained by injection molding machine Arburg 320 C type Allrounder 500–250 (screw 35 mm) from iPP HP551M (Basell, Ferrara; MF ≈ 8 g/10 mins). Processing conditions were injection velocity of 40 cm^3/s, melt temperature (nozzle) of 240°C, and injection pressure 1100 bar.

WNW fabrics with surface density of 80 g/m^2 were spunbonded by Texbond Spa (Rovereto, TN, Italy) by using iPP with melt flow of 18 g/10 mins (at 2.16 kg

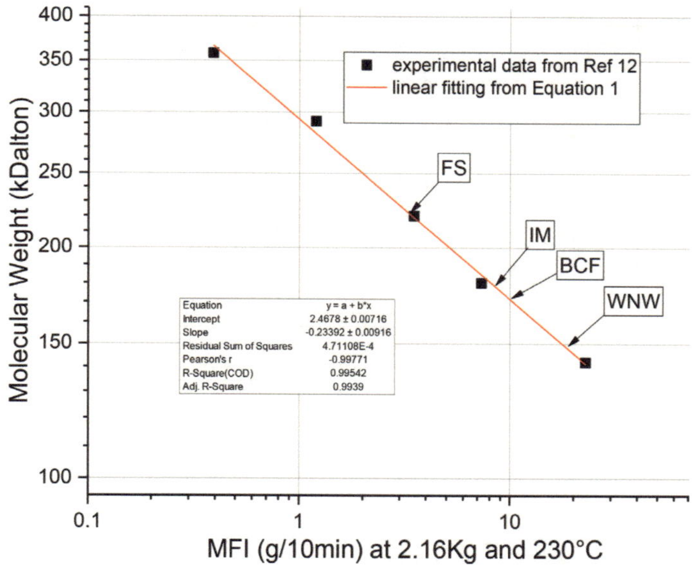

Figure 1.
Relation between melt flow index and molecular weight. The value of various products are indicated, i.e., woven non-woven (WNW), bulk continuous filaments (BCF), injection molding (IM), and fiber spinning (FS).

Code	Product	MF g/10 mins	Process
FS	Monofilament	3.6	Fiber spinning
IM	Dumbbell	8.5	Injection molding
WNW	Fabrics	18	Woven non-woven spunbonding
BCF	Multifilament	10.0	Bulk continuous filament spinning

Table 1.
Products obtained from polypropylene of different melt flow.

and 230°C). Due to the anisotropy of the process, fabrics were tested in machine direction (MD) and cross direction (CD) [13]. Normalized thickness of samples for mechanical testing was obtained assuming 0.905 g/cm as bulk density of polymer.

BCF of 64 filaments with the total titer of 1150 dtex [14] and equivalent diameter of 403 μm were industrially produced by Aquafil Spa (Arco, TN, Italy) by using PP with melt flow of 10 g/10 mins (2.16 kg, 230°C).

Monofilaments of PP and composite fiber were produced in lab-scale starting from commercial pellets (Sabic PP505P with MF = 3.6 g/10 mins at 2.16 kg and 230°C) and kaolinite masterbatch (Paralux by Vale, Brazil). Polypropylene fibers containing kaolinite in the range 1–30 wt% were manufactured by a two-step process, i.e., compounding/melt spinning and hot drawing. Melt-compounding was performed in a corotating intermeshing twin-screw extruder Rheomix Thermo Haake PTW16 (L/D 25; D = 16 mm; rod die 1.65 mm; temperature profile 130–230°C). Drawing of extruded filaments with 500 μm diameter was set at 145°C in order to produce single fiber at increasing draw ratio (DR) in the range 5–15. More details of compounding-spinning-drawing processes are described in literature [15, 16].

2.2 Mechanical and thermal tests

Tensile tests were performed on dumbbell ASTM specimens (Section 3.2 × 12.7 mm), BCF multifilaments (200 mm length), and WNW fabrics (50 mm width and 200 mm length) by using a dynamometer Instron mod. 4502 with a crosshead speed of 100 mm/min. WNW fabrics were tested both in machine and in cross direction. Single fibers of 10–20 mm were tested with a crosshead of 5–10 mm/min.

Differential scanning calorimetry was performed on PP specimens of about 15 mg by means of a Mettler DSC30 calorimeter by thermal cycling in the range 0–220°C with a heating/cooling rate of 10°C/min. Melting and crystallization temperature/peak were registered. The crystallinity X was determined referring the measured melting enthalpy ΔH_i in the first heating scan to 207 J/g and the standard enthalpy of the fully crystalline PP according to Eq. (2):

$$X = 100\ \Delta H_i/207. \qquad (2)$$

DMTA-thermal creep specimens of WNW (stripes 20x5 mm) and BCF (15 mm length) were subjected to dynamical mechanical analysis in tensile mode by using a DMTA Mk II (Polymer Laboratories) with a dynamic deformation of 11 μm, frequency of 5 Hz, static stress between 1 and 16 MPa, and heating rate of 3°C/min in the range −50/120°C. Storage (E') and loss (E") moduli are reported as a function of temperature. Thermal creep (TC) was also evaluated according to Eq. (3):

$$TC = 100^{*}(\Delta L/L_0) \qquad (3)$$

where ΔL is the specimen length variation and L_0 is the initial length.

Isothermal creep at 25°C was performed on samples 200 mm length (and 50 mm width for WNW fabrics) by applying for 30 mins a constant stress of 16 MPa for BCF and 3.5 MPa for WNW, and a following recovery for 30 mins at a minimum stress of 1.5 MPa for BCF and 0.05 MPa for WNW, respectively. WNW fabrics were tested both in machine and cross direction.

2.3 Fiber diffraction measurements

All diffraction images were collected in transmission using a modified Laue camera with a removable image plate 24 × 15 cm^2 with a pixel size of 43 microns at a distance from the sample equal to 8.81 cm. The sample to detector distance

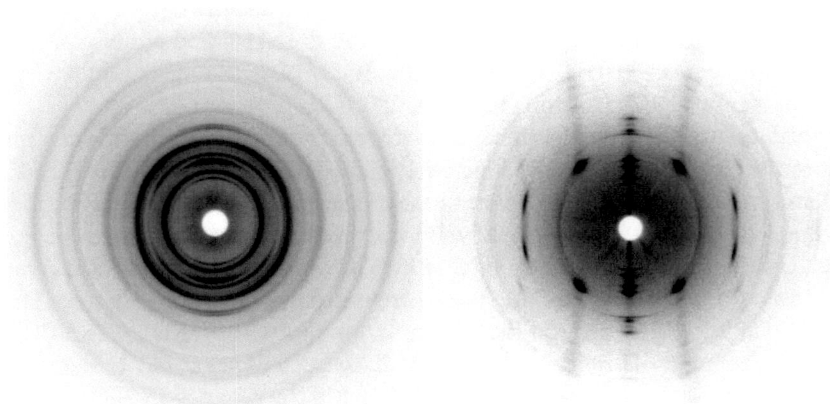

(a) (b)

Figure 2.
Fiber diffraction raw images for the PP fibers with 10 wt% kaolinite (K10). (a) As-span fibers showing smoother texture. (b) Fibers at DR = 10: the central spots radially dispersed are due to diffraction from the residual bremsstrahlung radiation in the filtered only X-ray beam. Sharper spots are produced by the PP strong fiber texture, and more continuous circles are from kaolinite diffraction.

was calibrated through a Si standard powder packed between two Mylar films and stretched by the same fiber diffraction sample holder used for the polymers.

The beam (CuKα radiation at 40 kV and 30 mA) was collimated through a pinhole and a Ni filter to ensure a proper resolution in the images and sufficient beam intensity. Only one diffraction image per sample was sufficient to get all the crystal, texture, and microstructure information needed.

In **Figure 2(a)** and **(b)**, two of these diffraction images are shown for the PP fibers with kaolinite filler. The two images enlighten how easily the differences in texture can be caught by the fiber diffraction technique. Not only the quality of the texture can be appreciated but also a quantitative analysis can be done by a proper methodology shown in the following paragraph. From the two images, we can also see some artifacts originated from the not strictly monochromatic beam (diffraction of the bremsstrahlung), but we account for them in our analysis.

3. Results and discussion

3.1 Mechanical properties

The effect of orientation of various polypropylene products can be firstly evaluated by the different mechanical properties; in particular modulus and maximum stress (strength) are summarized in **Figure 3**.

Dumbbell specimens exhibited a tensile modulus of 950 ± 9 MPa, yield stress of 33 ± 1 MPa, and stress and deformation at break of 15 ± 2 MPa and 73%, respectively. Mechanical properties are quite different from other products, even if molecular weight is quite similar to BCF products. In injection molding, chain alignment and solidification follow a different pattern with respect to the fiber formation during spinning, and consequently dumbbell specimens exhibit heterogeneity in macromolecular orientation with a skin effect and a disordered core structure, as well described in literature [11]. Moreover, it should be considered that the shape factor, calculated as the ratio between the perimeter and section, is lower for IM (0.8 mm^{-1}) with respect to 160 mm^{-1} for WNW, 80 mm^{-1} for BCF, and in the range of 8–31 mm^{-1} for FS depending on the drawing. The efficiency of chain orientation

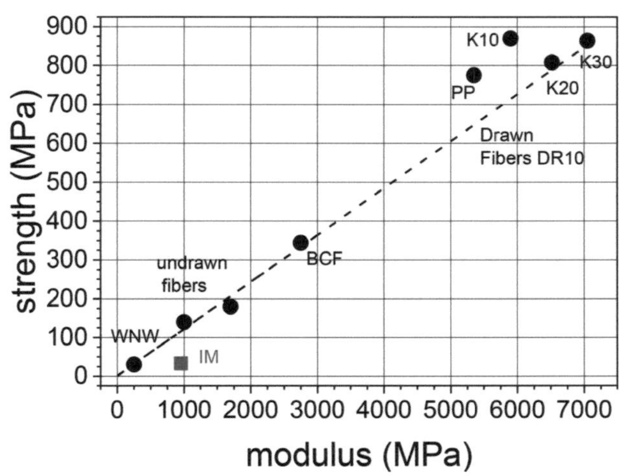

Figure 3.
Comparison of mechanical properties (tensile modulus and strength) of various oriented products, such as WNW, undrawn fibers from WNW, BCF, and drawn fibers with DR = 10; selected data of PP fibers or nanocomposite fibers with 10 (K10), 20 (K20), and 30% (K30) of kaolinite. Data of IM sample are also shown for comparison.

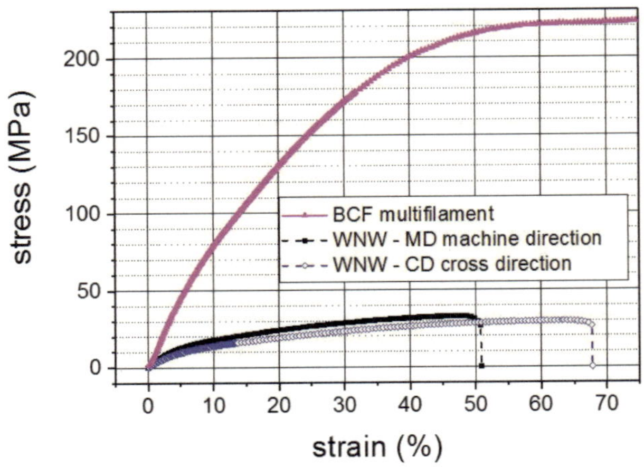

Figure 4.
Comparison of stress-strain curves of BCF filament and WNW fabrics tested in MD and CD directions.

is more evident in fiber-like products, where processing conditions determine a linear stretching (drawing) of polymer with elongation flow, either in spinning or in drawing.

The higher the orientation, the higher the modulus, the higher the strength, and in general the lower the strain at break. It is evident of the lower values of WNW fabrics with respect to undrawn fibers and BCF filaments. An almost linear dependence between strength and stiffness can be observed in **Figure 3**.

The highest modulus and strength have been obtained for lab-scale fiber after drawing 10 times (DR10) and relatively higher values for nanocomposites with kaolinite between 10 and 30%.

Stress-strain curves of WNW fabrics and BCF filaments are shown in **Figure 4**. The effect of fiber spinning in BCF and the direct drawing in the process with draw ratio of about 4 determines not only a stiffening of the filament but also an increase of max stress (320 MPa) and a relatively high deformation at failure (125%). More peculiar is the orientation in WNW fabric where longitudinal or machine direction and transversal or cross direction determine a different mechanical behavior. Anisotropy of WNW products is very common and usually decreases with the increase of surface density.

This effect is more evident in the creep curve shown in **Figure 5**, where CD sample exhibited not only a higher deformation after 30 mins creep with respect to MD sample (3.1 vs. 2.0%) but also a higher residual after 30 mins recovery (about 0.8 vs. 0.3%).

Correspondingly a higher storage modulus was found for MD sample with respect to CD sample, as shown in **Figure 6**. On the other hand the peak of the polymer glass transition temperature (Tg) remains localized at about 0°C.

The different thermal creep (at 3.5 MPa) of WNW with respect to BCF is reported in **Figure 7**, and it decreases in the order WNW-CD > WNW-MD > BCF. In all cases the deformation starts at about 30°C after the glass transition interval depicted by the loss modulus peak in the interval −15/30°C. As expected, the higher the orientation, the higher the storage modulus, the lower the loss modulus, and consequently the lower the thermal creep.

Figures 8 and **9** show DMTA data of BCF filaments. Storage modulus and thermal creep directly depend on the applied stress.

Moreover, it is evident that static stress of 1 or 7 MPa determines a thermal creep starting from about 30°C, as in the case of WNW with 3.5 MPa (see **Figure 7**). On the other hand, at higher stress of 16 MPa, thermal creep starts at 0°C that correspond to the Tg, measured at the maximum of loss modulus peak.

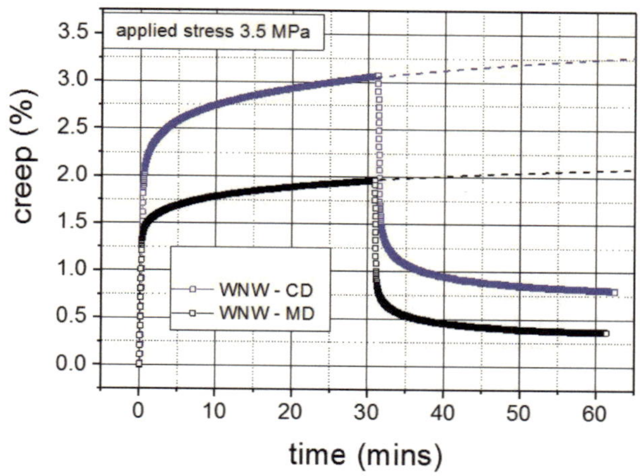

Figure 5.
Comparison of creep curves of WNW fabrics tested in MD and CD directions at 25°C with an applied stress of 3.5 MPa for 30 mins, followed by 30 mins recovery.

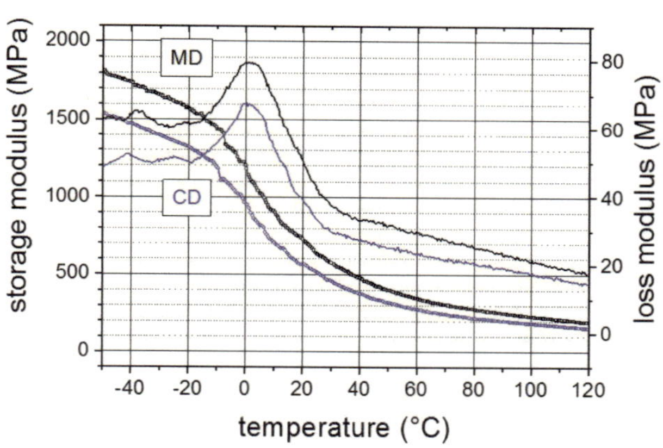

Figure 6.
Dynamic mechanical analysis of WNW fabrics performed at a static stress of 3.5 MPa. Storage modulus (□□□ ○○○) and loss modulus (——— ———) of MD and CD samples are compared.

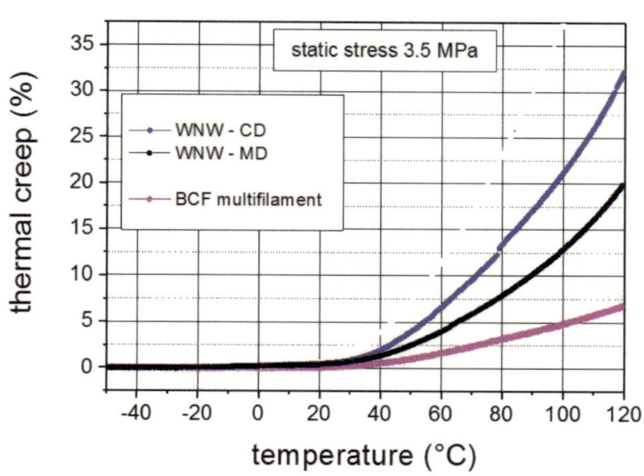

Figure 7.
Thermal creep comparison of WNW fabrics (MD and CD) and BCF filaments as measured by dynamic mechanical analysis with static stress of 3.5 MPa.

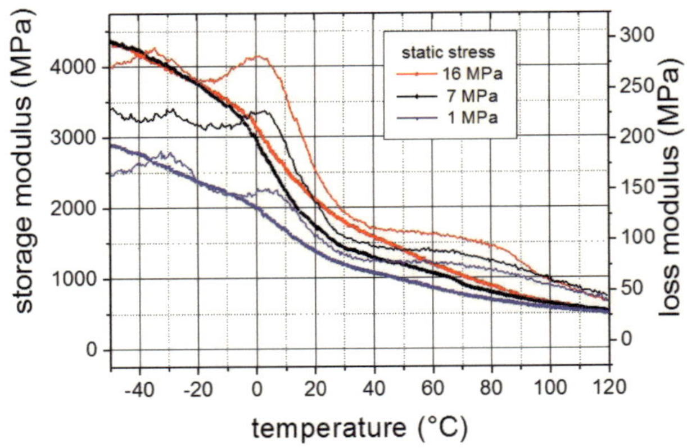

Figure 8.
Dynamic mechanical analysis of BCF filaments performed at different static stresses of 1, 7, and 16 MPa. Storage modulus (▪▪▪ ♦♦♦ ▴▴▴) and loss modulus (—— —— ——).

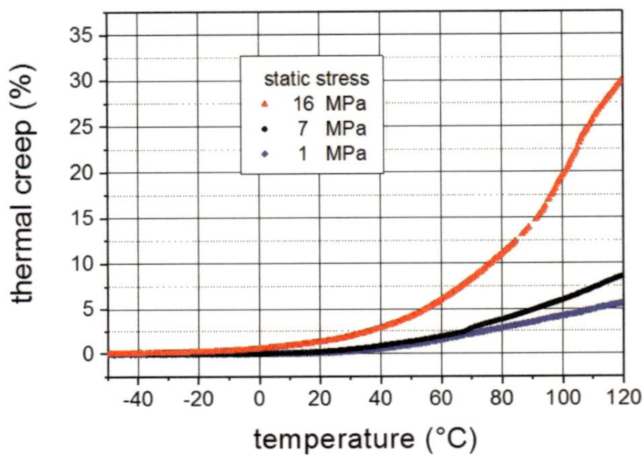

Figure 9.
Comparison of thermal creep measured in dynamic mechanical analysis of BCF filaments performed at different static stresses of 1, 7, and 16 MPa.

Creep curves of BCF filaments at 16 and 78 MPa are compared in **Figure 10**. Experimental data have been interpolated by using a simple exponential model, as previously described [16]. The two parameter K and n formally represent the intensity and the rate of the creep, and they are directly dependent on the applied stress.

Thermal analysis (DSC data) could be useful for a preliminary evaluation of samples. The first heating scan is representative of the products, whereas the cooling step and the second heating scan give information on the polymer. Crystallinity (endothermal peak), the quality of the peak from melting peak and crystallizability of polymer during the cooling step, usually depends on molecular weight. For IM sample melting temperature of 169.2°C and melting enthalpy of 83.2 J/g (crystallinity of 40.2%) were determined. **Tables 2** and **3** show selected data of WNW and BCF samples, respectively. DSC data of various WNW fabrics before and after testing are almost similar.

Thermal analysis of BCF original filaments and after creep or mechanical experiments revealed an interesting result of increased crystallization, in dependence on the well-known phenomenon described as "stress-induced crystallization" [17] or "orientation-induced crystallization" [18] that occurs either in fiber processing or

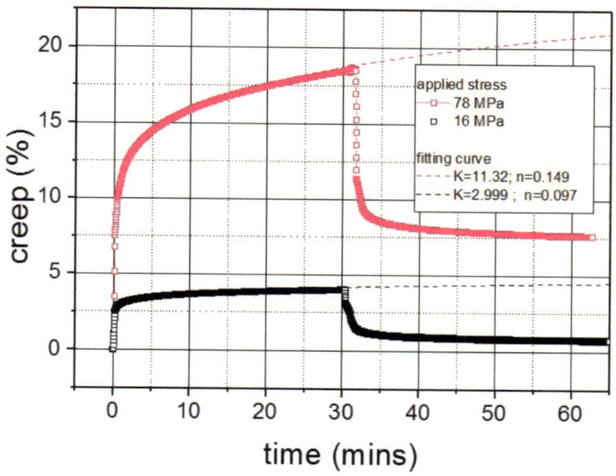

Figure 10.
Comparison of creep curves of BCF filament tested at 25.0°C with different high applied stresses of 11 or 78 MPa for 30 mins, followed by 30 mins of recovery.

WNW	First	Scan		Cooling		Second	Scan
	T_{m1} [°C]	ΔH_1 [J/g]	X [%]	T_c [°C]	ΔH_c [J/g]	T_{m2} [°C]	ΔH_2 [J/g]
As received	175.5	84.6	40.5	121.1	95.2	164.2	96.2
MD after creep 3.5 MPa	171.6	84.1	40.2	121.5	96.3	163.4	96.6
CD after creep 3.5 MPa	173.7	84.3	40.3	121.8	95.1	163.6	96.7
MD after failure	170.1	86.2	41.2	121.8	95.7	163.5	95.8
CD after failure	173.1	82.5	39.9	120.1	94.1	165.1	95.2

Crystallinity is calculated according to Eq. (2).

Table 2.
Thermal results of WNW fabrics (melting and crystallization temperature; melting and crystallization enthalpy in the three DSC scans).

BCF	First	Scan		Cooling		Second	Scan
	T_{m1} [°C]	ΔH_1 [J/g]	X [%]	T_c [°C]	ΔH_c [J/g]	T_{m2} [°C]	ΔH_2 [J/g]
As received	175.5	74.6	35.7	114.4	90.9	166.6	91.3
After creep 16 MPa	172.9	74.0	35.4	115.1	89.4	165.5	90.7
After creep 78 MPa	172.9	79.4	38.0	115.4	89.5	165.5	91.5
After failure	174.0	108.3	51.8	115.4	89.1	166.0	90.5

Melting and crystallization temperature; melting and crystallization enthalpy in the three DSC scans. Crystallinity is calculated according to Eq. (2).

Table 3.
Thermal results of BCF filaments.

in fiber testing. The residual deformation after creep and recovery was found at 0.7 and 7.5% after loading 16 or 78 MPa, respectively, and correspondently crystallinity of 35 and 38% was measured. It should be noted that crystallinity of BCF filaments increased up to 52% after the failure (strain failure at 125%).

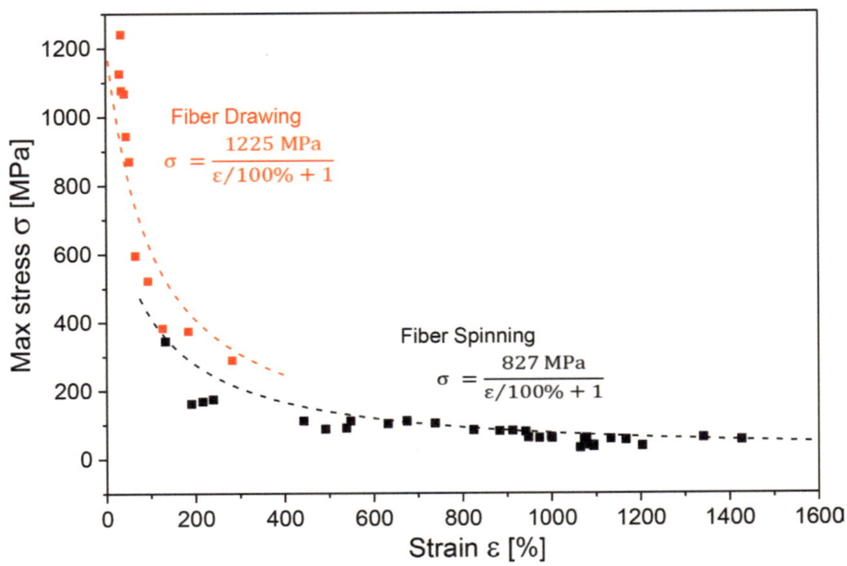

Figure 11.
Comparison of max stress and strain at failure of PP fibers in the two steps of processing, i.e., spinning and drawing. Interpolation curve following the equation proposed in literature [17].

3.2 Case of highly oriented products (fibers)

Single fibers at different levels of drawing were prepared and characterized. **Figure 11** shows the relationship between the max stress and deformation at break that could be interpolated according to the criteria of independence of a total maximum orientation or the network deformation concept [17]. It could be assumed as a direct combination of the orientation obtained in fiber spinning, the extension of the fiber during fiber drawing, and the strain deformation during tensile testing [17]. Experimental results of PP fibers have interpolated separately, distinguishing the spinning step and the drawing step.

The calculated values of 827 and 1225 MPa formally represent the maximum attainable strength in fiber spinning and fiber drawing. It is well evident that 1225 MPa is underestimated, suggesting that various and different phenomena occur during fiber production with different roles and consequences on the ultimate properties.

Moreover, the effect of the filler was also observed and evaluated in drawing, by comparing the modulus and the strength of the fiber as function of draw ratio, as shown in **Figure 12**. The higher the draw ratio, the higher the orientation, and the higher the modulus and the strength, for both PP fiber and composite fibers.

The maximum modulus and strength have been obtained after drawing at draw ratio of DR 15 for both iPP and composite fibers, with values in the range of 8–9 GPa and 900–990 MPa, respectively.

3.3 Texture analysis

Fiber diffraction using transmission images is a powerful technique to analyze polymers especially in fiber or textile form. From a just one diffraction image, it may be possible to get crystal structure information [19, 20], texture [21, 22], and also microstructural features [21]. A strong texture and crystallization may help the crystal structure solution and refinement the same way as the texture is used for crystal structure solutions [22]. The major problem in the quantitative analysis of these transmission images is to account for the texture in a correct way. Simple fiber

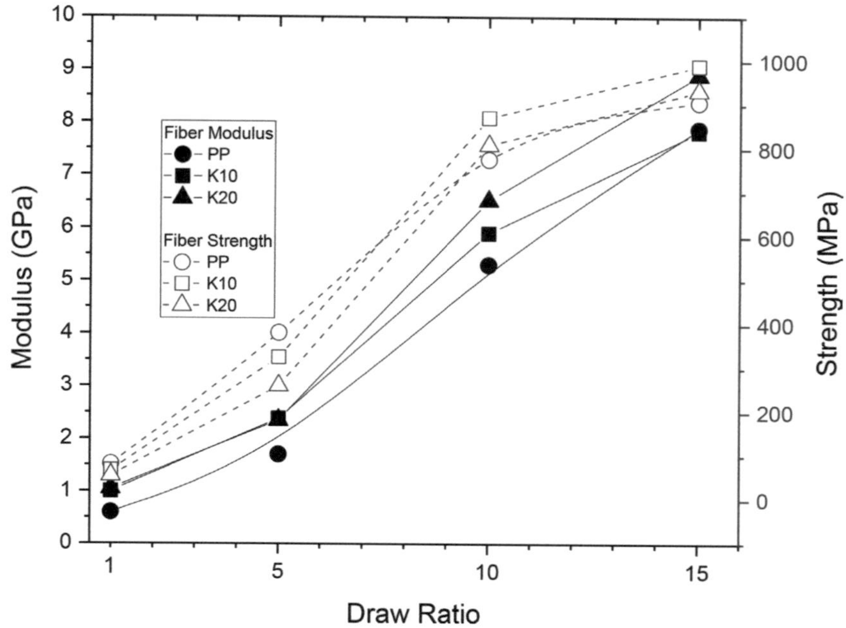

Figure 12.
Modulus and strength of PP fibers and nanocomposites with 10 and 20% of kaolinite as function of draw ratio.

textures can be modeled easily as proven by Ran et al. [23], and in-line analyses can be carried out to monitor the crystallizable behavior of polypropylene fibers. But only one attempt has been made to obtain some rough quantitative information on the texture by Jin et al. [24] for the polypropylene fibers. In this paper we will show a procedure to perform a global diffraction analysis from which we can obtain simultaneously all information from the crystal structure to the orientation distribution function (ODF). The ODF is a function describing the volumetric amount of material with a specific crystallographic orientation in one direction. The procedure is based on the Rietveld texture analysis [25, 26] but using transmission images [20, 21].

To briefly recall the general methodology, the 2D images collected in transmission are transformed in spectra sampling the image in radial slices each one covering a certain diffraction cone angle [27–29]. All the spectra are refined at once in the Rietveld refinement program MAUD [30] using in addition to the crystal structure and size-strain model a texture model to obtain the ODF. In the case of samples in fiber form, the more suitable model is the standard functions [20, 31] as it provides a sufficiently flexible way to represent the ODF with few refined parameters. Instead for smoother texture, like in the case of WNW samples, which we could not model with an ideal standard function, the more flexible spherical harmonic method [25, 31–33] has been used but in the exponential form to ensure a positive function.

Only isotactic polypropylene (iPP) was found in our samples, and for the fiber diffraction images fitting and refinement, we started from the crystal structure determined by Natta et al. [34, 35]. In order to reduce the number of degrees of freedom in our model, we used a bond and angle restraint function for the iPP. This was sufficient to drive our analysis to a unique solution and safely obtain the ODF that was our principal goal.

Figure 13 reports the fitting for the BCF as-span sample. The same analysis procedure was applied also to the strained fibers after creep/recovery at 78 MPa with residual deformation of 7.5% (as shown in **Figure 10**) and after thermal creep in DMTA at 16 MPa with residual deformation of 30% (see **Figure 9**). The experimental diffraction image on the left was transformed by unrolling around the

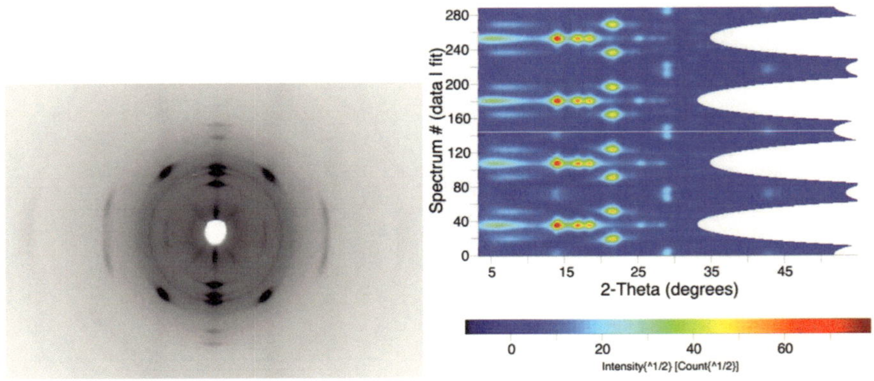

Figure 13.
Original fiber diffraction image for the BCF as-span on the left. Unrolled and fitted data at the end of the analysis on the right. The lower part of the right image (spectrum number from 0 to 143) contains the unrolled experimental data, on the upper part (spectrum number 144 to 287), the calculated patterns. The matching of the two parts indicates a good fitting and correct model.

diffraction center (the white hole) in steps of 2.5° in order to get 144 radial diffraction spectra that were fitted by the program MAUD. **Figure 13**, on the right, shows the result of the fitting. The calculated patterns in the top part well reproduce the unrolled experimental spectra in the bottom. The longer spots at low angles, and visible in the image on the right as radial short strings, are due to the diffraction of the bremsstrahlung that was also included in the pattern modeling. In the unrolled map on the right, the presence of sharper spots vertically means sharper texture. Instead, sharper spots horizontally correspond to a better crystallization of the fibers/compound.

In **Figure 14** we have recalculated some of the pole figures from the ODFs obtained for the three BCF samples. Pole figures, being 2D, are somehow more convenient to visualize the texture characteristics. All three different fibers show the same kind of texture: the fiber axis is parallel to the normal to the pole figures, and a perfect fiber symmetry was found for the (100) axis. For the texture analysis, we were obliged to use the monoclinic c-setting for the iPP instead of the more common b-setting. The usual (001) fiber axis becomes the (100) in our case, because of the different cell conventions used. The texture sharpness does not change significantly between the as-span and strained fibers at room temperature, but there is a strong increase in fiber alignment after thermal creep. In fact the fiber spread obtained from the fitting was 15 ± 1° for the as-span and 17 ± 1° for the room temperature creep but becomes 9 ± 1° for the fibers after creep in DMTA up to 120°C (see **Figure 9**). Also the mean crystallite sizes, as measured from the analysis, increase from 14 to 19 nm with the thermal creep, but it is not affected by the creep at room temperature. These findings are in agreement with the residual deformation of BCF filament, i.e., 30% after thermal creep (**Figure 9**) and 7.5% after creep/recovery (**Figure 10**).

For the WNM samples, for which we show only one experimental diffraction image and its fitting in **Figure 15**, the results of the texture analysis for the three samples, as received and after creep in MD and CD directions, are shown in **Figure 16**. The texture is weaker than in the previous BCF case and it is not a fiber one. The machine and cross directions show a slightly different fiber alignment that gives result to a different texture when subjected to creep in their respective direction. The texture sharpness increases a bit and more for the machine direction, as it was showing a more favorable alignment of the fibers from the beginning.

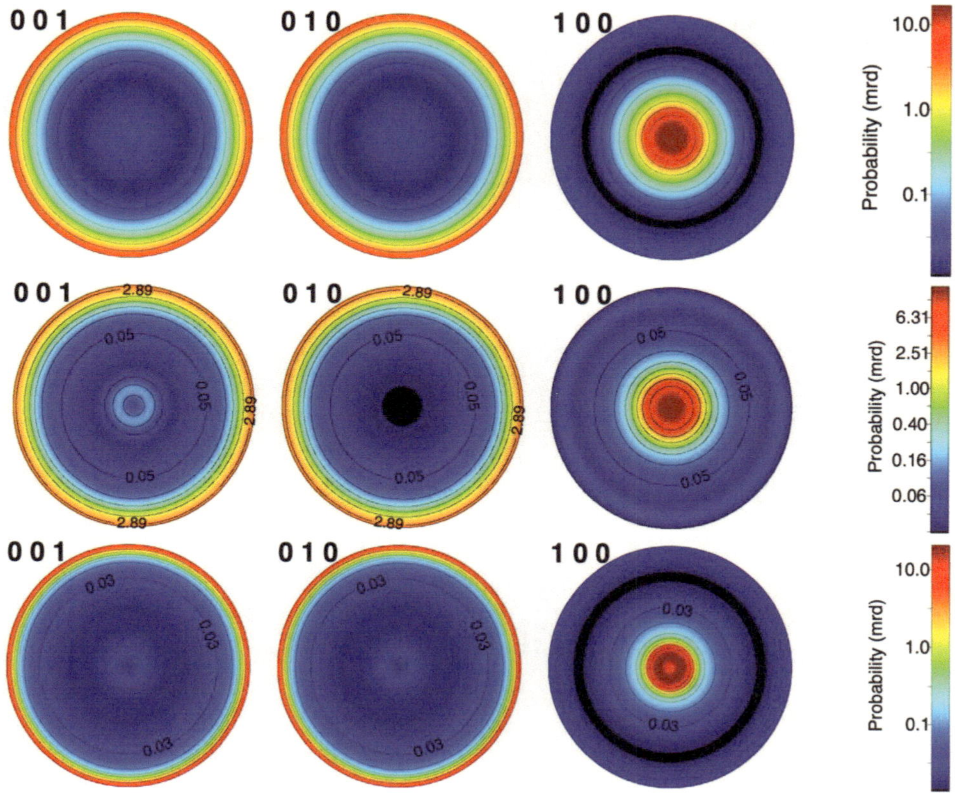

Figure 14.
Recalculated pole figures for BCF fibers (from top): as-span, after creep at 78 MPa (room temperature) and after thermal creep at 16 MPa. Only after creep at high temperature we notice an increase in the crystallographic alignment of the fibers. The fiber sample direction corresponds to the normal to the pole figures.

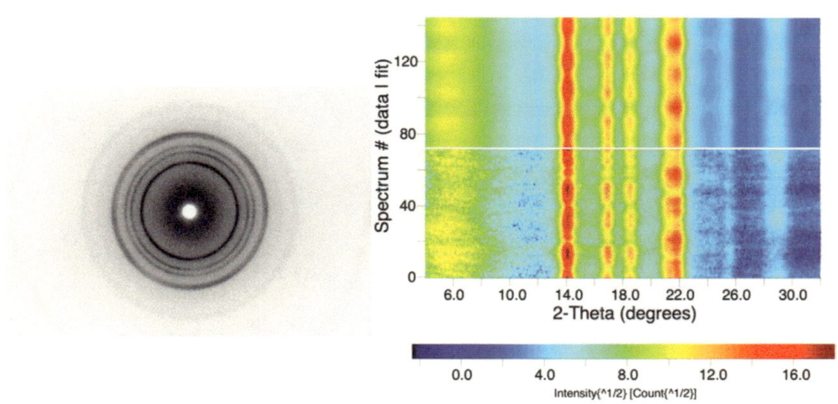

Figure 15.
Original fiber diffraction image (left) and unrolled and fitted patterns (right) for the WNW sample. The texture is much smoother with respect to the drawn fiber samples. With respect to **Figure 13***, the uncomplete 2Θ range has been cut to enhance low-angle features.*

Finally, we analyzed in fiber diffraction also one sample containing kaolinite. The analysis was more difficult in this case as two textured phases are present. Again, kaolinite shows a high density of modulated planar defects and strong compression in the plane normal to the fibers axis.

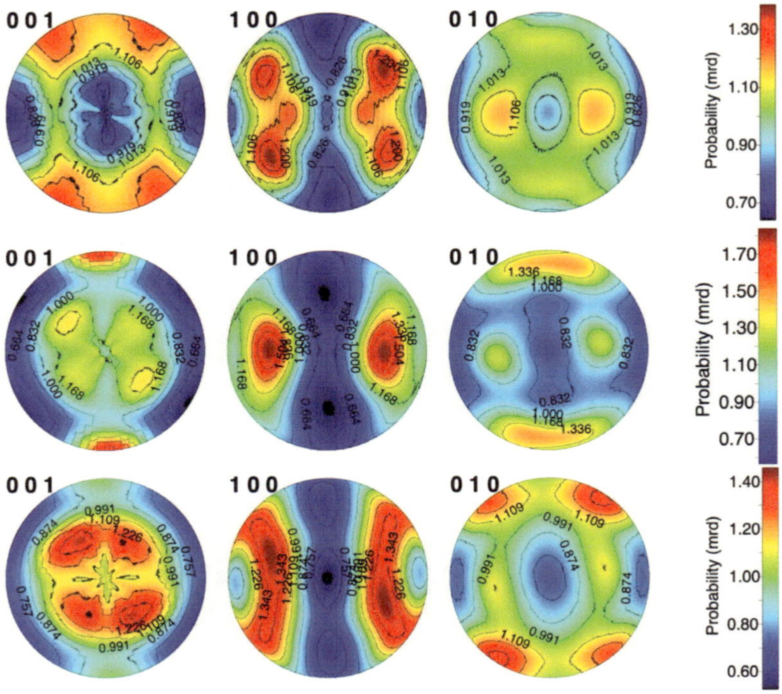

Figure 16.
Recalculated pole figures for WNW samples (from top): as prepared, after creep in machine direction, and after creep in cross direction. The pole figures' horizontal direction corresponds to the normal to the WNW in-plane tissue. The creep direction is normal to the pole figures for MD and vertical for CD.

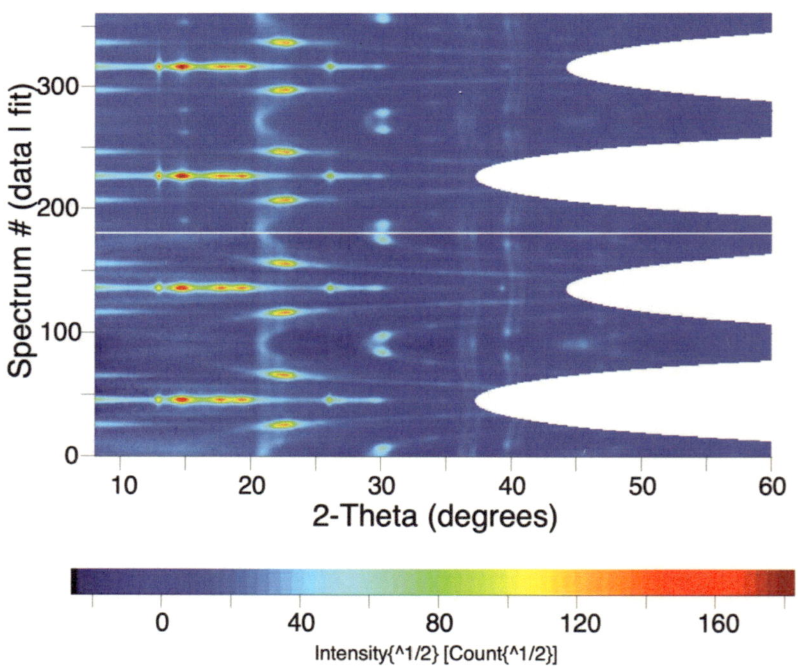

Figure 17.
Rietveld fitting of the unrolled diffraction image for the sample K10. The spectrum number corresponds to a pattern integrated radially every 2° in the circumferential direction starting from the horizontal plane. There are 180 experimental integrated patterns in the lower part and 180 recalculated patterns in the upper part. The waving vertical lines correspond to the kaolinite diffraction interplanar spacings in compression in the plane perpendicular to the fibers.

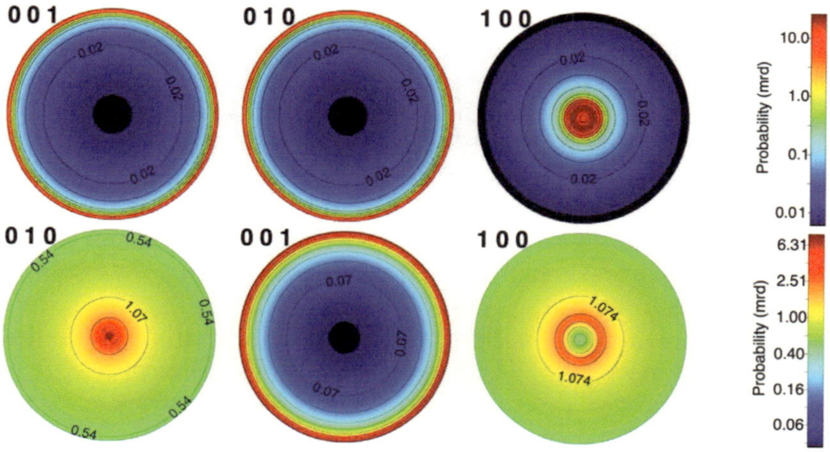

Figure 18.
Recalculated pole figures of iPP (top) and kaolinite (bottom) resulting from the Rietveld texture fitting. The fiber direction is normal to the pole figures.

To correctly model the unrolled diffraction patterns, as reported in **Figure 17**, we had to derive a modulated planar defect model for kaolinite starting from a previous turbostratic model [36]. This modulated defect structure has been tested by analyzing the powder diffraction pattern of the same kaolinite used as filler in the fibers.

The fitting of the fiber diffraction, for the iPP with 10 wt% kaolinite at DR = 10, is reported in **Figure 17**. The iPP texture and microstructure characteristics are similar to the BCF, especially the one after thermal creep, and by difference with **Figure 13**, the reader may recognize which ones are the diffraction spots and lines of kaolinite. Kaolinite shows a high compression state perpendicular to the fiber axis corresponding to about 0.8% equivalent elastic strain. This extremely high deformation is probably induced by the fiber drawing and the intercalation with kaolinite. By the strong texture, the deformation mainly belongs to the (001) kaolinite basal plane (perpendicular to the plane) that is also the intercalation and faulted plane.

From the crystallization point of view, the iPP is similar to the BCF after thermal creep (17 nm for the mean crystallite sizes), and kaolinite is arranged in packets of about 50 nm perpendicular to the basal plane.

Figure 18 is showing the recalculated pole figures for the iPP and kaolinite. The fiber spread of the iPP corresponds to 7.3 ± 0.5°, and so kaolinite contributes to the alignment of the fibers reaching a texture even sharper than the BCF after thermal creep. From the kaolinite pole figures, we can deduce that the (001) basal plane is distributed normal to the fiber axis but in a perfect fiber texture.

Finally, from the analysis of the fiber diffraction image, we can measure also the amount of kaolinite, and thanks to the proper modeling, including texture and strain effects, the refined amount was 10.7 wt% that is very close to the amount inserted (10 wt%).

4. Conclusions

It can be concluded that polymer processing directly affects mechanical properties of iPP fiber-like products, as observed in the case of WNW fabrics and fibers, both BCF and monofilaments. The higher the orientation, the higher the modulus and the strength, the lower the deformation at break, and the higher the creep resistance. No particular variation of crystallinity of specimen before and after creep and fracture test was detected by DSC analysis, except for BCF after failure.

In the case of composite fibers, the improvement in mechanical properties of monofilament is mainly dependent on the fiber drawing, whereas only a marginal contribute of kaolinite content has been observed. For instance, the processing-drawing with draw ratio of 10 produced monofilaments of iPP and kaolinite composite with modulus and strength in the range of 5.5–6.5 GPa and of 770–870 MPa, respectively.

From the texture analyzed by X-ray, we can notice that the fiber alignment is only affected by the creep at high temperature. Stretching the fibers at room texture has a negligible effect on the texture. The situation may be different for the WNW, where the fibers have a higher mobility and the stress can change their orientation, also at lower temperatures.

In addition, from this work we can conclude that the mechanical properties are highly correlated to the texture and reverse. The fibers after thermal creep, which are showing the higher residual deformation, show also a higher increase in the texture.

Acknowledgements

Authors wish to thank Aquafil SpA (Arco, TN, Italy) and Texbond Spa (Rovereto, TN, Italy) for providing BCF filaments and woven non-woven fabrics, respectively. Moreover, the authors acknowledge N. Soave for injection molding and I. Dabrowska, D. Lorenzi, and K.T. Chaka for fiber spinning.

This research received no external funding.

Conflict of interest

The authors declare no conflict of interest.

Dedication/other declarations

This work is dedicated to the memory of Prof. József Karger-Kocsis (†December 13, 2018).

Appendices and nomenclature

BCF	bulk continuous filament
CD	cross direction
DM(T)A	dynamic mechanical (thermal) analysis
DR	draw ratio
DSC	differential scanning calorimetry
FS	fiber spinning
IM	injection molding
MAUD	material analysis using diffraction, software for X-ray analysis
MD	machine direction
TEM	transmission electron microscopy
Tg	glass transition temperature
XRD	X-ray diffraction
WNW	woven non-woven fabric
ODF	orientation distribution function

Author details

Luca Fambri* and Luca Lutterotti
Department of Industrial Engineering, University of Trento, Italy

*Address all correspondence to: luca.fambri@unitn.it

References

[1] Kutsch O. Market Study: Polypropylene. 4th Ed. Available from: http://www.ceresana.com/en/marketstudies/plastics/polypropylene/ [Accessed: 18 October 2018]

[2] Plastics Europe. Plastics-The Facts 2018. An Analysis of European Plastics Production, Demand and Waste Data. 2018. 25 p. Available from: http://www.plasticseurope.org/download_file/force/2367/181 [Accessed: 13 December 2018]

[3] Moore EP. Polypropylene Handbook: Polymerization, Characterization, Properties, Processing, Applications. Munich: Hanser Publishers; 1996. pp. 303-348. ISBN: 3446181768

[4] Karian HG. Handbook of Polypropylene and Polypropylene Composites. New York: Marcel Dekker; 1999. pp. 15-37. ISBN: 0824719492

[5] Fambri L, Lorenzi D, Masarati E, Costantini E. Chapter 12: High load polypropylene composites. In: Silva LP, Barbosa EF, editors. Polypropylene: Synthesis, Applications and Environmental Concerns. Happauge New York: Novascience Publishers; 2013. pp. 261-284. ISBN: 978-1-62417-152-9. Available from: https://www.novapublishers.com/catalog/product_info.php?products_id=42212

[6] American Standard Testing Materials ASTM D1238-10: Standard Test Method for Melt Flow Rates of Thermoplastics by Extrusion Plastometer

[7] Tripathi D. Practical Guide to Polypropylene. Shawbury, UK: Rapra Technology Ltd; 2002. pp. 75-84. ISBN-10: 1859572820

[8] Wishman M, Hagler GE. Polypropylene fibers. In: Lewin P, Pearce EM, editors. Handbook of Fiber Science and Technology Fiber Chemistry. Vol. 4. New York: Marcel Dekker; 1985. pp. 371-497. ISBN: 0-8247-7335-7

[9] Mather RR. The structure of polyolefin fibres. In: Eichhorn SJ, Hearle JWS, Jaffe M, Kikutani T, editors. Handbook of Textile Fiber Structure. Vol. 1. Cambridge, UK: Woodhead Publishing; 2009. pp. 276-304. ISBN: 9781845693800

[10] Kunugi T. High-modulus and high-strength polypropylene fibers and films. In: Karger-Kocsis J, editor. Polypropylene an A-Z Reference. Dordrecht The Netherlands: Kluwer Academic Publishers; 1999. pp. 295-300. ISBN: 0412802007

[11] Fujiyama M. Higher order structure of injection-molded polypropylene. In: Karger-Kocsis J, editor. Polypropylene Structure, Blends and Composites. Vol. 1. London: Chapman & Hall; 1995. pp. 167-204. ISBN 0412802007

[12] Feng B, Fuming L, Calhoun BH, Quirk RP, Cheng SZD. Physical constants of poly(propylene). In: Brandrup J, Immergut EH, Grulke EA, editors. Polymer Handbook. 4th ed. New York: Wiley; 1999. p. V-21. ISBN 0-471-16628-6

[13] European Disposables and Nonwovens Association (EDANA) [Internet]. 1999. Available from: http:///www.edana.org/discover-nonwovens/how-they're-made/formation [Accessed: 30 November 2018]

[14] Bertamini L, Caldara M, Giacomelli G, Pontarin LM, Fambri L, Casagranda A. New fibres for floor-covering and textile applications. Chemical Fibers International. 2006;**1**:28-35

[15] Fambri L, Dabrowska I, Ceccato R, Pegoretti A. Effects of fumed silica

and draw ratio on nanocomposite polypropylene fibers. Polymers. 2017;**9**:41. DOI: 10.3390/polym9020041

[16] Dabrowska I, Fambri L, Batistella M, Lopez-Cuesta JM. Compounding and spinning of polypropylene nanocomposites with kaolinite. In: Proceedings of 16th Eur. Conf. On Comp. Materials (ECCM16); 22-26 June 2014; Seville, Spain; ID 725. 2014. pp. 1-8

[17] Beyreuther R, Brunig H, editors. Dynamics of Fibre Formation and Processing. Heidelberg: Springer; 2007. pp. 65-81. DOI: 10.1007/978-3-540-46223-1

[18] Jinan C, Kikutani T, Takaku A, Shimizu J. Nonisothermal orientation-induced crystallization in melt spinning of polypropylene. Journal of Applied Polymer Science. 1989;**37**(9):2683-2697. DOI: 10.1002/app.1989.070370919

[19] Chatani Y, Maruyama H, Noguchi K, Asanuma T, Shiomura T. Crystal structure of the planar zigzag form of syndiotactic polypropylene. Journal of Polymer Science Part C: Polymer Letters. 1990;**38**:393-398

[20] Lutterotti L, Bortolotti M, Fambri L. Crystal structure and texture refinement of polymers from diffraction images. Acta Crystallographica. 2005;**A61**:C391-C392

[21] Lutterotti L, Bortolotti M, Ischia G, Lonardelli I, Wenk HR. Rietveld texture analysis from diffraction images. Zeitschrift für Kristallographie Supplements. 2007;**26**:125-130

[22] Wessel T, Baerlocher C, McCusker LB. Single-crystal-like diffraction data from polycrystalline materials. Science. 1999;**284**:477-479

[23] Ran S, Zong X, Fang D, Hsiao BS, Chu B, Ross RJ. Novel image analysis of two-dimensional X-ray fiber diffraction patterns: Example of a polypropylene fiber drawing study. Journal of Applied Crystallography. 2000;**33**:1031-1036

[24] Jin Y, Rogunova M, Hiltner A, Baer E, Nowacki R, Galeski A, et al. Structure of polypropylene crystallized in confined nanolayers. Journal of Polymer Science Part B: Polymer Physics. 2004;**42**:3380-3396

[25] Ferrari M, Lutterotti L. Method of simultaneous determination of anisotropic residual stresses and texture by X-ray diffraction. Journal of Applied Physics. 1994;**76**(11):7246-7255

[26] Matthies S, Lutterotti L, Wenk HR. Advances in texture analysis from diffraction spectra. Journal of Applied Crystallography. 1997;**30**:31-42

[27] Lutterotti L, Matthies S, Wenk HR, Schultz AS, Richardson JW Jr. Combined texture and structure analysis of deformed limestone from neutron diffraction spectra. Journal of Applied Physics. 1997;**81**:594-600

[28] Lutterotti L, Vasin R, Wenk HR. Rietveld texture analysis from synchrotron diffraction images: I. Calibration and basic analysis. Powder Diffraction. 2014;**29**(1):76-84

[29] Wenk HR, Lutterotti L, Kaercher P, Kanitpanyacharoen W, Miyagi L, Vasin R. Rietveld texture analysis from synchrotron diffraction images: II. Complex multiphase materials and diamond anvil cell experiments. Powder Diffraction. 2014;**29**(3):220-232

[30] Lutterotti L. Total pattern fitting for the combined size-strain-stress-texture determination in thin film diffraction. Nuclear Instruments and Methods in Physics Research B. 2010;**268**:334-340

[31] Matthies S, Vinel GW, Helming K. Standard Distribution in Texture Analysis. Berlin: Akademie-Verlag; 1987

[32] Bunge HJ. Texture Analysis in Materials Science, Mathematical Methods. Berlin: Akademie-Verlag; 1969. ISBN: 0-408-10642-5

[33] Popa NC. Texture in Rietveld refinement. Journal of Applied Crystallography. 1992;**25**:611-616

[34] Natta G, Corradini P. Structure and properties of isotactic polypropylene. Nuovo Cimento, Supplemento. 1960;**15**:40-51

[35] Brückner S, Meille SV, Petraccone V, Pirozzi B. Polymorphism in isotactic polypropylene. Progress in Polymer Science. 1991;**16**:361-404

[36] Lutterotti L, Voltolini M, Wenk HR, Bandyopadhyay K, Vanorio T. Texture analysis of a turbostratically disordered Ca-montmorillonite. American Mineralogist. 2010;**95**(1):98-103